**Books are to be returned on or before
the last date below.**

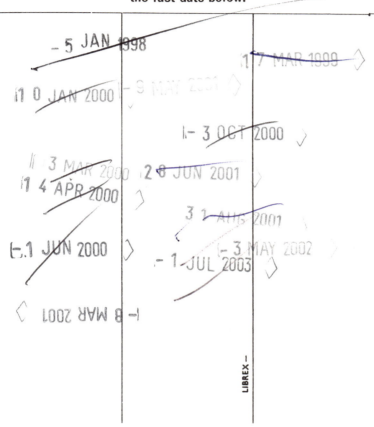

- 5 JAN 1998

17 MAR 1999

1 0 JAN 2000

- 9 MAY 2001

I- 3 OCT 2000

1 3 MAR 2000

2 8 JUN 2001

1 4 APR 2000

3 1 AUG 2001

- 1 JUN 2000

- 3 MAY 2002

- 1 JUL 2003

- 8 MAR 2001

LIBREX —

INTEGRATED M/E DESIGN

BUILDING SYSTEMS ENGINEERING

ANIL AHUJA

Chief engineer, Building Systems, Mechanical and Electrical, The Austin Company
Des Plains, IL

CHAPMAN & HALL

I(T)P® International Thomson Publishing

New York • Albany • Bonn • Boston • Cincinnati • Detroit • London • Madrid • Melbourne
Mexico City • Pacific Grove • Paris • San Francisco • Singapore • Tokyo • Toronto • Washington

For more information, contact:

Chapman & Hall
115 Fifth Avenue
New York, NY 10003

Chapman & Hall
2-6 Boundary Row
London SE1 8HN
England

Thomas Nelson Australia
102 Dodds Street
South Melbourne, 3205
Victoria, Australia

Chapman & Hall GmbH
Postfach 100 263
D-69442 Weinheim
Germany

International Thomson Editores
Campos Eliseos 385, Piso 7
Col. Polanco
11560 Mexico D. F.
Mexico

International Thomson Publishing - Japan
Hirakawacho-cho Kyowa Building, 3F
1-2-1 Hirakawacho-cho
Chiyoda-ku, 102 Tokyo
Japan

International Thomson Publishing Asia
221 Henderson Road #05-10
Henderson Building
Singapore 0315

Table 3.3 and figures 3.2-3.7 were reprinted from IEEE Std 141-1993 IEEE Recommended Practice for Electric Power Distribution for Industrial Plants, Copyright ©1994 by the Institute of Electrical and Electronics Engineers, Inc. The IEEE disclaims any responsibility or liability resulting from the placement and use in this publication. Information is reprinted with the permission of IEEE.

1 2 3 4 5 6 7 8 9 10 XXX 01 00 99 98 97
Library of Congress Cataloging-in-Publication Data

Ahuja, Anil.
    Integrated mechanical/electrical design : building systems
engineering / Anil Ahuja.
        p.   cm.
    Includes index.
    ISBN 0-412-09831-8
    1. Buildings--Mechanical equipment. 2. Buildings--Electrical
equipment 3. Systems engineering.   I . Title.
TH6010.A58   1996
696--dc20   96-26709
                                                  CIP

British Library Cataloguing in Pubication Data available
"Integrated M/E Design: Building Systems Engineering" is intended to present technically accurate and authoritative information from highlyl regarded sources. The publisher, editors, authors, advisors, and contributors have made every reasonable effort to ensure the accuracy of the information, but cannot assume responsibility for the accuracy of all information or for the   nsequences of its use.

To order this or any other Chapman & Hall book, please contact **International Thomson Publishing,
7625 Empire Drive, Florence, KY 41042.** Phone: (606) 525-6600 or 1-800-842-3636.
Fax: (606) 525-7778. e-mail: order@chaphall.com.

For a complete listing of Chapman & Hall's titles, send your request to **Chapman & Hall, Dept. BC,
115 Fifth Avenue, New York, NY 10003.**

Contents

Preface

Buildings are constructed to serve the needs of their occupants. Occupants need facilities with a comfortable, safe, healthy environment; utilities, communication and technical equipment to perform their work. Building systems provide and support security, comfort, utilities and technical needs.

A building system may be defined as a group of electro-mechanical components connected by suitable pathways for the transmission of energy, materials or information and directed to a specific purpose. Energy may take any form, from the old fashioned air, water and steam pressure to gas and electricity. If a system is to provide optimum performance, it's individual components should be well matched, and their contributions to overall system performance should be clearly defined by the designer and understood by the operator. The integrated systems approach gives consideration to the overall objectives rather than the individual elements, components and subsystems.

Traditionally, mechanical and electrical engineering coordination in buildings is primarily between mechanical and electrical systems. In the past two decades the integration of diverse technologies such as: mechanical, electrical, bio-climatology, geophysics, optics, electronics and computer engineering play important role in the design of building systems and the environment they control. Building systems is not a curriculum discipline in its own and in detail engineering of an individual discipline it cannot be covered like in a single discipline textbook, but in each discipline essential points can be discussed in a basic form. Until now, there has been no single, consolidated source for the general information of all building system elements and issues provided in this book, though the need for it has been evident.

The computerization and integration of building system technologies and the information age we live in have changed the way humans perceive their habitat. Illustration of all business, health, energy, atmosphere, ground and engineering issues related to buildings and it's environment is one of the distinguishing features of this book. The systems approach is addressed in a language that non-engineers can understand and comfortable analogies are provided throughout to reduce it's complexity and the specialized impression. This is important because today's information oriented society and business internationalization demand quarterly profits, and also in turn, demands optimum efficiency and maximum reliability of building support systems at minimal cost to the owner. The design of integrated reliable building system is important because of high costs associated with system shutdown.

As with so many of the subjects in this book, the complexity of the building systems contrasts sharply with the relative simplicity of the basic mechanisms. This book provides an initial look at underlying mechanisms that keeps the reader from getting lost in specific intricacies. Another objective of this book is to present the fundamentals of integrated system reliability as it applies to planning and design of building systems. The text material is primarily directed toward facility and building architects, engineers, owners, investors and financiers.

The book provides a unique look into how humans, who are made of about 70 percent water and 30 percent solids, respond to the ways air and water moves in their electrified living habitats and compares that to the movement inside their body. The body is not isolated from the outside world and has a dynamic system openly exchanging with the outside environment. The body's functions, like buildings, are directed at maintaining correct temperature and humidity, receiving adequate mix of clean air and water molecules and relieving waste hygienically to render reliable performance up to seventy-five plus years.

This cross currents approach to understanding synergy of body and building systems is shaped by concepts of human bio-technology and building technology. Not only does this new way to vision buildings and its systems broaden our understanding of our own internal system operations , but it also relates to the factors of electricity and magnetism in our electrodynamic environment. After all, we build structures on Earth that operate on Earth's radiation fields, use Earth's gravity fields, survive in Earth's magnetic field and communicate through network of man-made electromagnetic fields. This can be visioned as Humans seeking buildings to shield from fields within natural fields within cosmic fields.

Most American companies face tremendous pressure to reduce utility cost, improve efficiency, provide safe and healthy atmosphere for their workers and meet environmental protection agency requirements. To achieve all this, a systems approach to integrate a buildings isolated operation is very effective. Highly integrated systems share information through networks, reduce system size and complexity, and allow the system to interoperate at peak performance. The book shows how standard protocols enables easy information exchange and create buildings that use micro technology for complete environmental control, illumination and sound control, energy demand reduction and safety.

Building systems reliable operation is directly linked to reliable production and minimum downtime. Reliability is usually measured in terms of time between failures for old systems and time of failure for new systems. This book helps managers and engineers to focus on real issues of risks, reliability, availability, and failure probability of a building system.

In the world of limited economic resources building owners , financiers and designers are seeking the most effective life cycle cost design solutions. The emphasis of this book is to provide owners a guideline to understand utility billing, integrate energy units and resources, include all system elements to avoid economic analysis errors, spot problem areas and reduce costs.

Current building system issues such as dirty air, power and water, their causes and solutions are discussed in detail. Topics like solar power, geothermal, thermal and electro-chemical storage are explained in simple language and help owners design buildings that are nature propelled and pays marketplace dividends.

The most important purpose of this book is to make the intelligent building systems engineering challenge more inviting.

1

Building Systems Engineering

1.1 An Introduction to Building Systems Engineering

The field of building systems engineering forms a cornerstone in modern building systems. Building systems engineering involves integration of such diverse technologies as civil, architectural, mechanical, electrical, electronic, and computer engineering. Engineering education has traditionally concentrated around single discipline activities. Mechanical engineering degree courses, for example, have generally included some aspects of electrical and electronic technology, and, similarly, electrical engineering courses have usually incorporated some aspects of mechanical engineering. The reason for this is that both of these engineering schools have been patently aware of other requirements and have at least recognized the need for an appreciation of the other discipline.

Building system applications include heating, ventilation, air conditioning, and cooling (HVAC); lighting; power; security; fire and life safety; building automation; and intelligence, data, video, and audio communications of various kinds. Figure 1.1 shows a small commercial building with mechanical and electrical systems typically found in a building. The majority of industrial enterprises involved with the manufacture of building systems products require engineers who have skills both in electrical and mechanical engineering. They also now require their engineers to be competent in computer technology and software concepts. Computers are mathematical instruments of enormous value in the *science* of building system design. Further, one of the graphics capabilities of computers is drafting; computer-aided drafting is an extremely valuable tool in the *art* of building system design.

Most of our academic engineering study is of the science, and it requires disciplined study and must be learned in sequence. One cannot pick up a little electronics here and some little mechanics there and accumulate either understanding

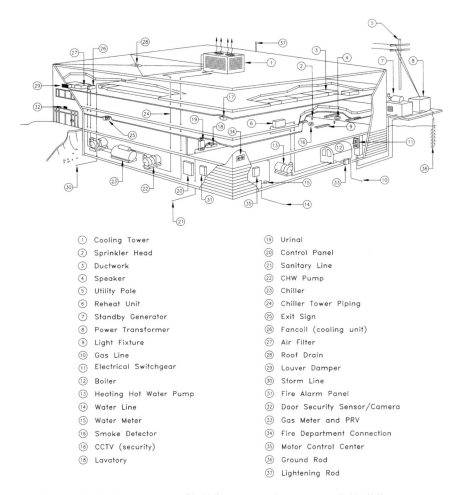

① Cooling Tower	⑲ Urinal
② Sprinkler Head	⑳ Control Panel
③ Ductwork	㉑ Sanitary Line
④ Speaker	㉒ CHW Pump
⑤ Utility Pole	㉓ Chiller
⑥ Reheat Unit	㉔ Chiller Tower Piping
⑦ Standby Generator	㉕ Exit Sign
⑧ Power Transformer	㉖ Fancoil (cooling unit)
⑨ Light Fixture	㉗ Air Filter
⑩ Gas Line	㉘ Roof Drain
⑪ Electrical Switchgear	㉙ Louver Damper
⑫ Boiler	㉚ Storm Line
⑬ Heating Hot Water Pump	㉛ Fire Alarm Panel
⑭ Water Line	㉜ Door Security Sensor/Camera
⑮ Water Meter	㉝ Gas Meter and PRV
⑯ Smoke Detector	㉞ Fire Department Connection
⑯ CCTV (security)	㉟ Motor Control Center
⑱ Lavatory	㊱ Ground Rod
	㊲ Lightening Rod

Figure 1.1 Typical components of building systems in a commercial building.

or utility. And using cookbook or code book formulas might be thought of as a black box type of approach, which nonetheless is perfectly viable. There are many other aspects of design called follow approaches that are not subject to logic at all. Consider *robustness, aesthetics, and customers' and managers' tastes and prejudices,* which are few of many.

The *science* of building system engineering design is mathematical analysis of devices and systems. Engineers use the science to *predict performance* and to *size parameters.* The *art* of engineering design is the knowledge of everything else that can be useful in design and the skill to use that knowledge. The art *can* be piecemeal and can go on throughout life. A vital portion of the art of design engineering is a knowledge of building technology and economics.

The building electromechanical engineer needs a working familiarity with the operating characteristics, performance, and practical applications of all building systems. These systems might include complex natural phenomena and mechanisms; diverse hydraulic, thermal, hydronic, and chemical processes; integrated electrical and auxiliary system operations; or any combination of these. A firm grasp of the basic underlying principles is the key to success for the practicing engineer. An intuitive understanding of the behavior of building systems, matter, and energy enables integrated design engineers to imagine and understand the behavior of existing and proposed building systems to a degree that those untutored in the integration of systems cannot match.

It is not particularly surprising that mechanical engineers use an electrical analogy to describe many heat conduction phenomena, while electrical engineers often use heat and fluid flow analogies to enhance an understanding of electrical and electronics principles. Analogies such as electrical voltage to mechanical pressure, electrical resistance to mechanical friction, electric wire to mechanical pipe, and electric switch to mechanical valves are key to better understanding, but should be used with caution. It is convenient to establish an analogy between electric and mechanical systems as an aid to comprehension but it should not be carried so deeply as to lose its parallel relation.

For example, electric current is a measure of power flow and, as such, would correspond to water flow in a hydraulic system. The correspondence is not complete, however, since in the hydraulic system the velocity of water flow varies, whereas in the electric system the velocity of propagation is constant. A break in electrical circuit stops the flow of current entirely (voltage is still present but no current can flow), whereas in a hydraulic pipe break the flow may actually increase until the pump fails or the supply of fluid is exhausted. Similarly, electrical switches are not quite like mechanical valves because they don't leak like a valve and larger pipes carrying more water flow are not analogous to bigger wire carrying more current, because what limits the allowable current through any wire is how hot the wire gets.

It is clear that every engineer, of whatever discipline, involved in building design will be faced with the need to understand the operation of building equipment and systems in the practice of other disciplines. Further, all building system activities need instrumentation and control equipment that is largely electrical, but monitors and controls mechanical equipment. Conversely, mechanical engineers will need to understand motors and motor drives that control and operate their machines. Electrical and mechanical engineers should be particularly concerned with developing system techniques.

This new breed of multidisciplinary engineer is at the heart of current building systems technological developments that continue to exhibit ever higher levels of sophistication and digital-based intelligence within the design of modern building electromechanical systems. This book assumes that electrical and mechanical engineers know their particular subject in depth but develops the interrelationship between ambient environment and all individual systems in building systems engineering.

1.2 The Emergence of Building Systems Engineering—Historical Perspective

Throughout history, the development of more sophisticated tools has often been associated with a decrease in dependence on human physical energy as a source of effort. Generally, this is accomplished by control of nonhuman sources of energy in an automated fashion. The building system revolution represented a major thrust in this direction.

A historical path that architecture has followed begins from simple shelters, well adapted to their climate. With the emphasis on energy control, architecture progressed to enclosed, tightly controlled internal climate structures that strive to ignore the conditions outside. The accumulation of energy conservation measures in the buildings resulted in poor air quality in some buildings, created power quality problems, and labeled some buildings as suffering from "sick building syndrome." Present architecture boasts of "intelligent" buildings, equipped with microcomputers and integrated systems—praised for their low annual energy consumption, clean indoor air, and superior power quality.

The basic transistor was invented in 1948, but it was not until the development of the microprocessor that building systems design engineers had taken recourse to a much more active involvement in the applications and utilization of microprocessor and integrated circuit devices in intelligent building electromechanical design. The trend in modern construction, except for small or simple structures, is clearly to use integrated system design with centralized monitoring and control of building systems. The subsystems in the 1980s almost always included in a building control system were HVAC, energy management, and lighting control. Inclusion of security, life safety (fire alarm, fire control and suppression, plus emergency aspects of vertical transportation), material handling, maintenance management, data/audio/video communications, and some aspects of office automation is the trend of future. The integration between mechanical, electrical, and computer technology has since become known as mechatronics and it is now becoming recognized as a curriculum topic in its own right.

A new awareness of the sources, characteristics, and limitation of energy supplies is resulting in new directions in building design, away from many practices of the recent past. This changing pattern in the use of energy sources has been accompanied by a trend in design: integrating building systems so that building's external skin and internal organization work with the surrounding climate for "natural" (passive) heating, cooling, and lighting. In most cases, a new equipment or system integration makes it possible to perform a familiar task in a somewhat new and different way, typically with enhanced efficiency and effectiveness and sometimes with increased understandability as well.

When building systems became so interrelated that it was no longer possible for a single individual to design them, and a design team was then necessary, a host of new problems emerged. This is the situation today. To cope with this, a number of methodologies associated with systems design engineering have evolved. Through these it has been possible to decompose large design issues into smaller component subsystem design issues, design the subsystem, and then build the complete system as a collection of these subsystems. Even so, problems remain. Just simply connecting together the individual subsystems often does not result in a system that performs acceptably, either from an efficiency or from an effectiveness perspective. This has led to the realization that systems integration engineering and systems management throughout an entire system life cycle are necessary.

The purpose of building systems integration is to design, install, and operate the many systems that comprise a facility so that their performances are not counterproductive. The output of an individual system should not jeopardize the performance, protection, and reliability of other systems.

1.3   Basic Vocabulary of Building Systems

Modern building support systems often bring information and signals beside power and gas utilities and process information data and signals at one place similar to producing heating and cooling in one place. The building system distribution structure then distributes them to other building spaces according to their respective needs. A substantial portion of building system design is dealing with the following 4 Cs to distribute all this flow:

Cessation is achieved with barriers, insulators, breakers, doors, curtains, glass, and switches.

Constraint is achieved through resistors, inductors, capacitors, orifices, pipes, stainers, filters, valves, fittings, insulation, connectors, conductors, and ducts.

Conduction of flow is achieved through use of conduits/pipes/ducts, conductors, vanes, windows, and doors.

Control of flow is achieved through valves, protectors, controllers, transducers, sensors, remote signals. What flow? Here is a list:

- Gas/vapor/air
- Heating/cooling/fire protection/domestic-liquid
- Sewage/sanitary/storm
- Electric current/magnetic flux

- Microorganisms/odors/pollens/ions/dirt
- Humans/heat
- Noise/vibration
- Information/light.

An integrated system designer has to have a vocabulary of other discipline specified materials, equipment, accessories, product features, and assemblies required for integration beside distribution. The system designer selects and combines the items in his vocabulary into the building system's design. Expanding your building system design vocabulary is a lifetime program.

See Figures 1.2 and 1.3 for integrated building system components. A building system designer's vocabulary should as a minimum include:

Energy and Fuel Services
Fuel service, storage, handling, piping, and distribution
Electrical service entrance and distribution equipment
Gas service, metering, and distribution piping

Heat-Producing Equipment
Boilers and furnaces
Steam-water converters
Heat pumps or resistance heaters
Make up air heaters
Heat-producing equipment auxiliaries

Power Producing Equipment
Generators, batteries, fuel cells
Photovoltaics

Refrigeration Equipment
Compressors, chillers, or absorption units
Cooling towers, condensers, well water supplies
Refrigeration equipment auxiliaries

Heat Distribution Equipment
Pumps, reducing valves, piping, piping insulation, etc.
Terminal units or devices

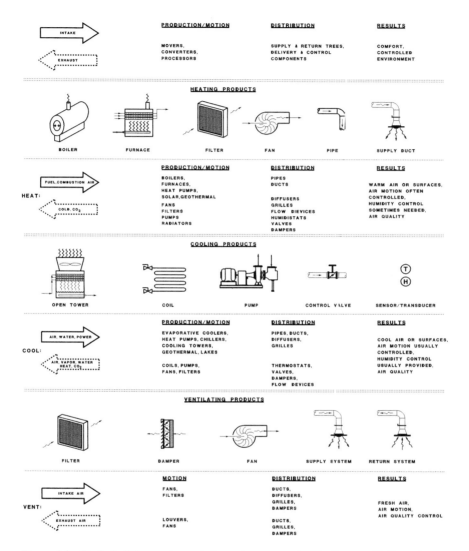

Figure 1.2 Basic building systems: tasks and components. Heating, ventilating, air conditioning (HVAC).

Cooling Distribution Equipment

Pumps, piping, piping insulation, condensate drains, etc.

Terminal units, mixing boxes, diffusers, grilles, etc.

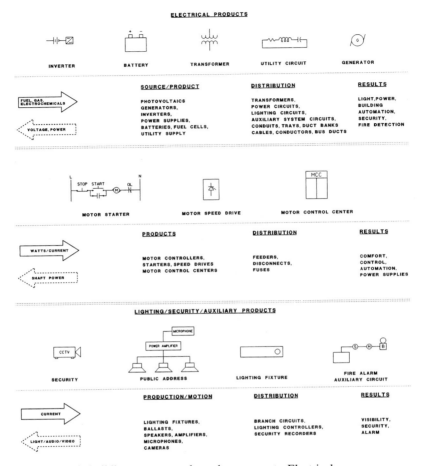

Figure 1.3 Basic building systems: tasks and components. Electrical.

Air Treatment and Distribution Equipment

Air heaters, humidifiers, dehumidifiers, filters, etc.

Fans, ducts, duct insulation, dampers, etc.

Exhaust and return systems

Air quality control equipment

Power Distribution Equipment

Transformers, switchgear, panelboards

Protective devices, device rating, and coordination

Feeders, starters, drives, auxiliary power equipment

Power conditioners, Power supplies, lightning, and surge protection equipment

Lighting Equipment

Lighting fixtures, reflectors, ballasts, daylighting

Lighting sources (HID, fluorescent, incandescent)

Lighting controllers, occupancy sensors, dimmers

Fire Protection Equipment

Standpipe and hose system

Sprinkler system—wet, dry, deluge, or preaction

Dry chemical system

Foam system

Auxiliary System Equipment

Audio/video and intelligence communication equipment and media

Fiber-optic connectors, receivers, and transmitters

Security system equipment, sound masking, and paging system equipment.

System and Controls Automation

Terminal or zone controls

System program control

System integrators, programmable controllers, network

Protocol, alarms, and indicator system

Building Site Resources, Envelope, and Environmental Systems

Passive or radiative heating and cooling equipment

Windows, glazing, and daylighting system

Solar and geothermal heating/cooling systems

Building Construction and Alteration

Mechanical and electric space

Chimneys and flues

Building insulation

Solar radiation controls

Acoustical and vibration treatment

Distribution shafts, machinery foundations, furring

Some basic nonelectrical engineers need to know: What allows electricity to flow? and How can electricity make a magnet? Electricity flashes through copper wire but halts when it bumps up against rubber tubing. Such examples abound in nature—certain substances are good conductors, while others block even the weakest electric current. Whether a material transmits electricity is dictated by the material's atomic structure. How easily a material channels electricity depends on how freely electrons move through it.

Conductors—chiefly metals such as iron, nickel, silver, and copper—have "loose," or free, electrons. Not leashed to any particular atom, these free electrons roam through their atomic neighborhoods, orbiting first one atom and then another. When a conductor is connected to a battery, the electric field organizes the aimless movement of the electrons into a steady flow. For this reason, metals are excellent carriers of electricity. Insulators, however, have few, if any, free electrons. Atoms in materials such as leather, glass, plastics, and rubber keep a tight rein on their electrons. An absence of freewheeling charges keeps insulators from conducting current.

A semiconductor is a crystalline material that conducts electricity but not as well as metals; it also resists electricity but not as well as many insulators. In general, semiconductors' electrons are tightly bound to their nuclei. But if a few atoms of antimony—with a surplus of electrons—are incorporated in a semiconductor such as silicon, the free electrons will give it a negative charge. These properties make semiconductors useful in transistors to amplify current, to block current, or to let current flow in only one direction. In a typical NPN transistor, a layer of positive (P) semiconductor, the base, is sandwiched between two negative (N) layers, the emitter and the collector. When, for example, a small signal from an intercom is channeled through the base, the movement of electrons amplifies the signal.

The discovery that electric currents generate magnetic fields led scientists to develop a magnet using electricity that can be turned on and off. Such electromagnets can consist of a battery attached to a coiled wire—a solenoid—wrapped around a ferromagnetic core (usually iron). The magnetic field produced by the electric current in the wire magnetizes the piece of iron. As long as current flows through the wire, the electromagnet behaves like a permanent magnet.

Magnetic field lines arc from the electromagnet's north to its south pole—usually at right angles to the flow of the current, in keeping with electromagnetic laws. If the current switches direction, the magnet's poles flip and the field lines reverse as well. The overall shape of the magnetic field does not change, however. The pattern of field lines remains the same unless the shape of the wire itself changes. Motors, generators, and other electrical systems operate by electromagnetism.

2

Atmospheric Radiation Soup and Buildings

2.1 Introduction

The daylight and the warmth we feel on a fine summer's day reach us as solar radiation. As everyone has known for many thousands of years, it is the sun that makes life possible and that sustains it continually. The radiation we see and feel, however, represents only a small part of the total radiation emitted by the sun.

The sun is a "main sequence" star. This sounds as though our local star is rather ordinary. It is, but that is not what the term means. Stars spend the greater part of their active "lives" burning hydrogen which is why this is called their "main sequence." The sun is a main sequence star because it is still burning hydrogen. So far, it has burned about half of its available stock—not all the hydrogen can be burned—and its present composition is about 90 percent hydrogen, 8 percent helium, and 2 percent consists of heavier elements. This energy released by the burning of hydrogen makes the center of the sun very hot—the temperature is calculated to be about 21.6 million °F (12 million °C).

2.2 Basics of Radiation

Electromagnetic radiation consists of a stream of photons. Photons are particles. They carry no electrical charge and it appears that they have no mass. You may wonder how it is possible for a particle to exist and yet to have no mass. The answer calls for the suspension of our common sense view of the world, which is necessary when considering the world of subatomic particles. A particle may have energy because it possesses mass and also because it is in motion. If a photon is stopped, however, it is destroyed and disappears, but it moves faster than any other particle. It is this that leads physicists to conclude that its energy consists only of the energy of its motion—it has no mass.

All photons travel at the same speed, the speed of light—in a vacuum, this is about 186,000 miles per second (300,000 km/s). Not all photons have the same amount of energy, however. Because a photon has no mass and it can move at only constant speed, there is only one way in which the difference in energy between one photon and another can appear—in its wave characteristics. The greater the energy a photon has, the shorter will be the distance between one of its wave peaks and the next. Its energy can be expressed as its wavelength. Conversely, the closer together the waves, the greater the number of wave peaks that will pass a fixed point in a given time.

The number of wave peaks per second is called the "frequency" of the wave and, therefore, the relationship between the wavelength and frequency of any regular wave motion is fixed, and the energy of the photon can also be expressed as its wave frequency. The wavelength is equal to the speed of propagation of the wave (in the case of electromagnetic radiation, the speed of light) divided by the frequency, and the frequency is equal to the speed divided by the wavelength.

The unit of frequency is the hertz (Hz), 1 Hz being equal to one wave cycle per second. This unit is often used to describe radio waves. A brief look into the history of the Hertz: Until approximately 1946, frequency was measured in cycles per second or just cycles for short. After World War II, the United States recruited German engineers and scientists who had developed Hitler's V1 and V2 weapons and brought them to the United States to work on rocket programs. The Germans found that although most electrical units were named after famous scientists (volt after the Italian scientist Volta, ampere after the French scientist, Ampere, and so forth), oscillation frequency was named in simple English. So the Germans declared that cycles per second should be called hertz after the famous German scientist Hertz, and they sold the idea.

The range of electromagnetic wavelengths is called its "spectrum" and it is divided into two parts: ionizing and nonionizing radiation. The ionizing radiations have sufficient energy to break down atoms of other materials, causing electrons to be dislodged and leaving atoms positively charged or ionized. Ionizing radiation consist of x-rays, gamma rays, cosmic rays, and other rays with frequencies above 10^{15} Hz. The part of the solar spectrum related to building systems operation—mostly nonionizing with ultraviolet at a frequency of 10^{15} Hz on the upper level proceeding down to visible light, infrared, microwaves, radiofrequency, and power frequencies at 50–60 Hz—is quite narrow, but it is the region in which the sun radiates most intensely. Because the spectrum is so large, different parts of it have been given their own common names (Figures 2.1 and 2.2).

There are several ways to describe the energy of a particular part of the electromagnetic spectrum, each with its own unit. The joule (J) is the scientific unit of energy and the electronvolt (eV) relates the energy of the photons to that of an electron. Conventionally written as one word, one eV is the energy needed to

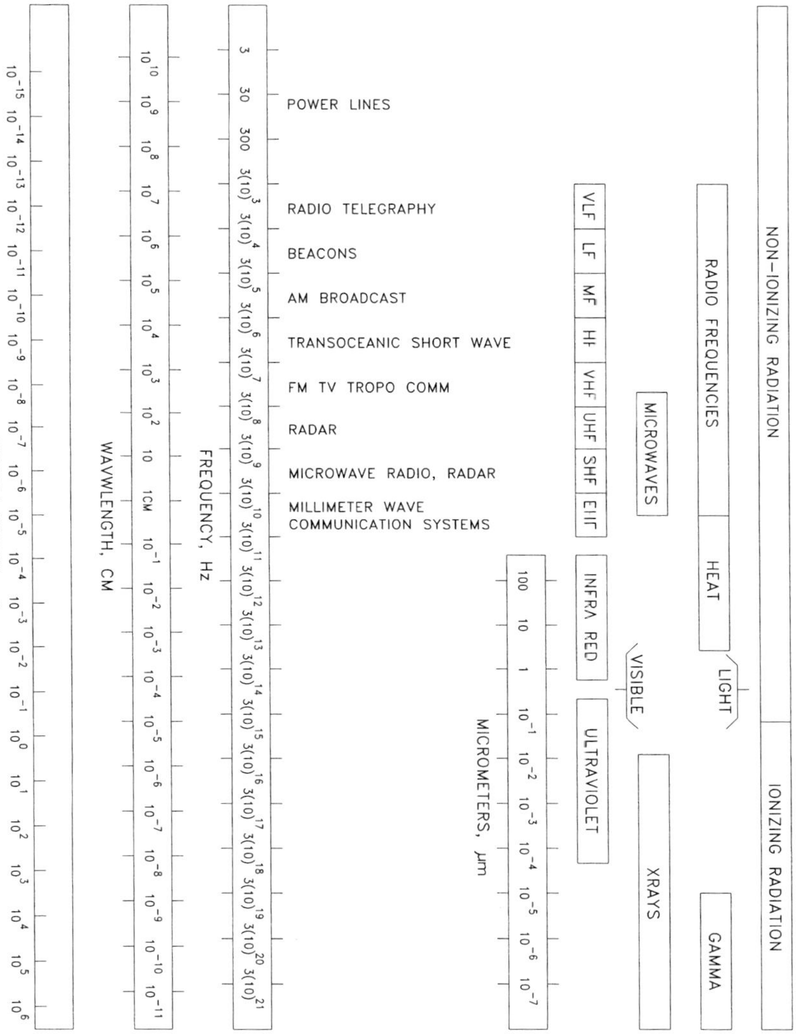

Figure 2.1 Frequency spectrum.

Electromagnetic spectrum
The top logarithmic scale gives frequencies in hertz; the lower one gives equivalent wavelength measurements. The lower the frequency the longer is the wavelength; the higher the frequency the shorter the wavelength will be. Hertz measurements are used for low frequency waves, and wavelength measurements for higher frequency waves. Sections of the spectrum are located between the scales, and illustrated with examples indicating their use or character.

A) Generated electricity
The alternating current (AC) electricity used in some countries for domestic and commercial use is of the low frequency/long wavelength end of the spectrum, and ranges from 16.7–144kHz. Two common frequencies: 50Hz (Europe) and 60Hz (North America). Two lines (1,2) on diamond 'A' indicate two common frequencies: 50Hz (Europe) and 60Hz (North America). These are heard as the 'mains hum' from faulty domestic gadgetry, hi-fi systems, etc.

B) Induction heating
Waves in this frequency band are used to induce heat in metals. The metal is placed in the center of a series of wire coils through which current is passed, inducing electrical eddy currents in the metal, which rise its temperature. Commonly the frequency of the current supplied to the coils is in the range of 60–60,000Hz, but waves with frequencies up to 500,000Hz are used.

C) Radio waves
These are the waves used for radio and television transmission. They range from 3kHz to 30GHz and, as shown on diamond 'C', may be subdivided as follows:
1. Very low frequency (vlf).
2. Low frequency (lf) waves, used for ship radio signals.
3. Medium frequency (mf), as used by police forces.
4. High frequency (hf), used for shortwave radio.
5. Very high frequency (vhf), used for radio and television.
6. Ultra high frequency (uhf).
7. Super high frequency (shf).

D) Microwaves
Waves in this region are used in radar (radio detecting and ranging), a method of detecting otherwise invisible objects by bouncing radio pulses off them. Such pulses are transmitted along a carrier wave in the range 1–35GHz (diamond 'D', region '1'). The shorter wavelengths in the microwave region, down to 1mm (diamond 'D', region '2'), have found application in cooking, greatly reducing cooking times.

E) Infrared waves
These are heat waves, with wavelengths between about 1mm and 7700Å. They are subdivided into far, middle, and near infrared ('1', '2', '3' on diamond 'E'). Most of the energy radiated by hot objects (including the energy we receive from Sun) lies chiefly in this range. Practical applications are in areas of photography (plates that are sensitive to infrared radiation make it possible to take photographs in the dark), physical therapy, military reconnaissance, and astronomical research.

F) Light waves
Although scientists use it more widely, the term "light" is most commonly taken to refer to the electromagnetic spectrum, ie to wavelengths of approximately 3900–7700Å. The color of an object is determined by the composition of its surface, which reflects certain wavelengths but not others. White objects reflect all wavelengths of the visible part of the spectrum; black objects reflect none of them.

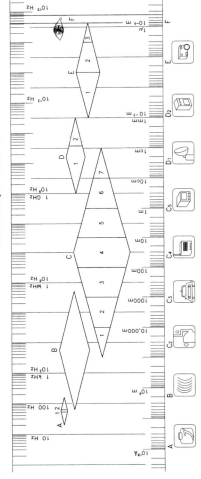

Figure 2.2 Building system electromagnetic waves.

14

move one electron through a potential difference of one volt. More commonly, the spectrum is divided into sections according to the wavelength or frequency of the radiation. Sunlight consists of a range of wavelengths each visible as a distinct color, but which we see as white when they are mixed together. Literally, these are the colors of the rainbow, because the water droplets that break sunlight into the bands of a rainbow are separating it according to its wavelengths.

2.3 Protective Role of the Atmosphere

The atmosphere is the gaseous envelope of the earth. All gasses in the atmosphere exist in a temperature equilibrium that is energized and maintai d in balance by the sun. The earth intercepts only about 0.002 percent e total amount of radiation emitted by the sun. This amount is proportional to the distance between the earth and the sun and, at the average distance of about 93,200,000 miles (150,000,000 km), it comes to about 1131 watts per square yard (1353 W/m^2). This is the amount of radiation that would reach the surface if none of it deflected, reflected, or absorbed on its way. It is called the "solar constant," although variations in the output from the sun mean it is not quite so constant as people once believed and, of course, it is an average—it increases as the earth approaches perihelion and decreases as it approaches aphelion.

The figure for the solar constant seems low until you remember that the sun radiates in all directions and the earth is a very small target. Most of the solar output comes nowhere near our planet. Of the solar radiation that does reach us, the solar constant, about 51 percent—an average of 595 watts per square yard (690 W/m^2)—penetrates all the way to the surface and about 19 percent—215 watts per square yard (257 W/m^2)—is absorbed in the atmosphere and so contributes to the total amount of energy available to us from the sun.

The proportion of sunlight that reaches the surface when the sky is cloudy depends on the extent of the cloud cover, but also on the type and thickness of the clouds themselves. Estimates of the extent of cloud cover are included in all reports from meteorological stations. The figure is usually expressed in "oktas"— one okta is equal to one-eight of the sky—but sometimes you may see it reported in tenths or as a percent age. The effect on the local radiation budget is calculated by deducting from the incoming radiation an amount determined by the type of cloud and the number of oktas.

Visibility is more severely reduced around dawn and sunset, because the sun is lower in the sky and, therefore, its more oblique radiation must pass through a greater thickness of mist and more of it is lost. It is evident, therefore, that, despite their reflective brilliance, clouds are not opaque. Light can and does penetrate them and inside a cloud, but near the top, the light can be very bright. This is the reason pilots are taught that, when flying in a cloud, they must rely wholly on their instruments to judge the altitude of the aircraft, because their ordinary senses become very unreliable.

2.4 Radiation Balance

Incoming and outgoing energy can be compared like the credit and debit columns of a balance sheet. A calculation of this kind produces a "radiation budget" that, like a financial budget, must balance. It can also be expressed as a "heat budget." Units of radiant energy can be translated into heat units because all the incoming energy that is absorbed—by the atmosphere as well as at the surface—is converted into heat. "Absorption" means that energy is delivered to the atoms and molecules of the receiving substance, raising their temperatures. Even photochemical processes cause energy to be released as heat or, in the case of photosynthesis, to be stored for later release as heat.

The balances do not have to be achieved immediately or locally. There can be, and are, imbalances according to season and latitude. In high latitudes, for example, the ground surface loses more energy over a year than it receives and, in low latitudes, where the sun is more directly overhead, the ground receives more energy than it loses.

There can also be very local variations. A hillside that faces away from the equator may never be warmed by direct sunshine and may lose more energy than it receives, while the other, equator-facing, side of the same hill receives more than it loses. Elsewhere, there are seasonal imbalances, with gains exceeding losses during the summer and losses exceeding gains during the winter. Even in a particular place and during a particular season, the balance can change from one hour to the next when clouds shade the surface. It is only over a number of years and over the planet as a whole that the totals must balance if the world is not to become hotter or colder.

The sun behaves as an almost perfect "black body." That is to say, it radiates energy very efficiently, at an intensity related to its surface temperature of about 11,000 °F (6000 °C). If the earth were also an efficient black body, promptly radiating back into space all of the energy it receives, the average surface temperature on the planet would be about −9 °F (−23 °C). The earth would be locked into a permanent ice age more severe than any it has ever experienced. Clearly, this is not the average temperature on the surface of the earth; over the whole surface and a full year it is about 59 °F (15 °C).

The wavelength at which a black body radiates is determined by its temperature, which is exactly what you would expect. The wavelength of radiation represents the amount of energy it conveys, and so the higher the temperature of the radiating body, the more energetic the radiation it emits will be. At 59 °F (15 °C) the earth radiates very slightly at less than 0.0004 mm, but mainly between 0.004 mm and about 0.04 mm, and most intensely at around 0.01 mm. The short-wave limit, of 0.004 mm, is taken as the boundary between "short-wave" and "long-wave" radiation.

Where the temperature is higher—over the subtropical deserts or during subtropical summer, for example—the wavelength is toward the shorter end of the wavelength, and where the temperature is lower—over high latitudes—it is toward the longer end. It is the difference in the amount of radiation absorbed in different latitudes that produces our climates, as heat is transferred from warmer to cooler regions through the atmosphere and oceans.

The amount of energy involved is considerable, at least by human standards. A commonplace local shower of rain involves about the same amount of energy as is required to keep a modern airliner airborne for 24 hours, and some local thunderstorms release as much energy as the burning of 7000 tons of coal. "Powering" the Asian monsoon requires as much energy as used by all the factories, offices, cars, and homes in the world.

The earth emits long-wave radiation but this does not leave the planet immediately. Atmospheric carbon dioxide, water vapor, and ozone absorb almost all radiation at wavelengths below 0.008 mm to 0.012 mm. At about 0.01 mm, however, which is where the earth radiates most intensely, there are gaps— "windows"—between the absorption bands of water vapor and ozone and between those of ozone and carbon dioxide at about 0.015 mm. Especially when the air is fairly dry, long-wave radiation escapes through these windows.

Of each 100 units of energy arriving at the top of the atmosphere, 34 are reflected. The atmosphere absorbs 19 units and only 47 units are absorbed by earth's surface. See Table 2.1 for debits and credits on the radiation balance sheet. The radiation that is absorbed in the atmosphere, together with latent heat released by the condensation of water vapor, is also reradiated in the same way and in all directions. Some escapes directly into space, some is directed downwards, and some is directed to the sides, where it is absorbed and radiated again or scattered in all directions by atmospheric molecules and particles. See Figure 2.3. It is because of the absorption of long-wave radiation in the atmosphere— the "greenhouse effect"—that the average temperature on the earth is held above its theoretical black-body value.

2.5   Buildings and Radiation Albedo

The word "albedo" is derived from the Latin *albus,* meaning "white" and, in a sense, it is a numerical measure of whiteness or, more correctly, of the proportion of radiation a surface reflects. In summer, many of us wear white clothes, or at least light-colored ones and, in winter, we tend to favor darker colors. We believe that white clothes reflect heat and dark clothes absorb it. This is true, although the number of layers of clothes and the materials from which they are made is a more important factor in keeping us warm or cool than their color.

Table 2.1   Surface plus Atmosphere Energy Balance

Incoming solar radiation (100%)	
Absorption by O_3	3
Absorption by H_2O and aerosols	13
Absorption by clouds	3
Backscatter by clouds	25
Backscatter by air and aerosols	7
Backscatter by the surface	2
Total	53
Absorption by the ground	
From the sun	25
From the atmosphere and clouds	22
Total	47
Total incoming radiation	100
Outgoing terrestrial radiation	
Short-wave	
Back-radiation from H_2O, CO_2, and clouds to outer space	32
Back-radiation from ground to outer space	2
Total	34
Long-wave	
Back-radiation from ground	6
Back-radiation from H_2O, CO_2, and clouds	60
Total	66
Total outgoing radiation	100

Within regions, albedos change from time to time. Snow cover in winter increases albedo and the spring thaw reduces it again. As you would expect, fresh snow is the surface with the highest albedo. Depending on how smooth and clean it is, its value lies between 0.8 and 0.9—between 80 and 90 percent of the radiation falling upon it is reflected. As the snow melts, its albedo falls to between 0.4 and 0.6. Sand comes next in the ranking, with an albedo of 0.30 to 0.35—desert sands tending toward the higher value.

Bare rocks have an albedo of 0.12 to 0.18. The albedo of urban areas is the same as that of rocks. Buildings and roads are made from rock, after all. The color of a building surface has such a strong influence on the proportion of solar radiation that is reflected and absorbed that it is a major factor in determining the size and capacity of building systems. An increase in its albedo means a surface absorbs less radiation. This will lower its temperature and that of the air above it, which, in turn, will reduce the rate of evaporation of water and cloud formation. Less cloud, however, will partly compensate by allowing more radiation to reach the surface.

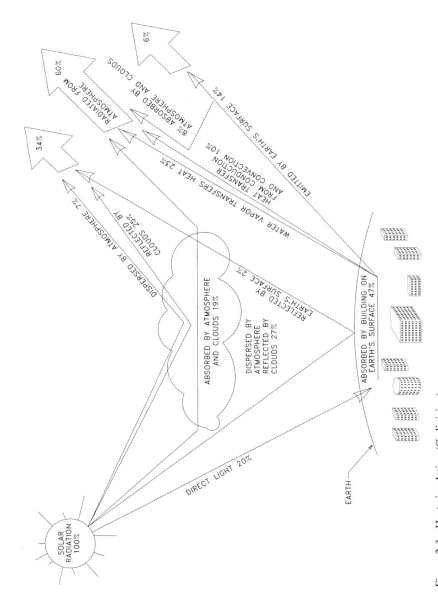

SOLAR RADIATION 100%

DIRECT LIGHT 20%

DISPERSED BY ATMOSPHERE 7%

DISPERSED BY ATMOSPHERE REFLECTED BY CLOUDS 25%

REFLECTED BY CLOUDS 27%

ABSORBED BY ATMOSPHERE AND CLOUDS 19%

REFLECTED BY EARTH'S SURFACE 2%

ABSORBED BY BUILDING ON EARTH'S SURFACE 47%

EARTH

34%

60%

6%

RADIATED FROM ATMOSPHERE

8% ABSORBED BY ATMOSPHERE AND CLOUDS

EMITTED BY EARTH'S SURFACE 14%

HEAT TRANSFER FROM CONDUCTION AND CONVECTION 10%

WATER VAPOR TRANSFERS HEAT 23%

Figure 2.3 Heat circulation (% division).

19

Some clouds that is, say, 70 feet (21 m) thick will reflect about 40 percent of the incoming radiation (its albedo is 0.40). A cloud 200 feet (61 m) thick has an albedo of about 0.65, one 400 feet (122 m) thick 0.78, and one 600 feet (183 m) thick 0.80. Very deep clouds, such as cumulonimbus storm clouds, may extend for 10,000 feet (3000 m) or more and have an albedo of 0.90.

Such high albedo values suggest that cloud cover cools the ground below, but matters are not quite so simple. Apart from warming the surrounding air by absorbing a small amount of incoming radiation and by the latent heat of condensation, water vapor absorbs long-wave radiation. This radiation, from the surface, is prevented from escaping into space and so clouds have a warming as well as a cooling effect. During the day, when the intensity of incoming radiation exceeds that of outgoing radiation, clouds block incoming radiation and have a cooling effect. At night, when incoming radiation ceases, they block outgoing radiation and so have a warming effect. Their overall influence is to make hot days cooler and cold nights warmer.

2.6 Ozone Layer

Near the *stratopause* of stratosphere zone at a height of about 30 miles (50 km), solar radiation at the short-wave end of the ultraviolet (UV) waveband splits oxygen molecule into their constituent atoms and the UV energy is absorbed by the atoms. The atoms rejoin to form molecules, some of them in threes to form ozone, and ozone also absorbs UV radiation, the atoms of its three molecules separating. The repeated breaking and reforming of oxygen and ozone molecules blocks virtually all UV radiation at wavelengths below about 0.0004 mm. Mixing of air in the stratosphere carries some upper air, enriched in ozone, to lower levels where the ozone is below the main area of UV absorption and accumulates, especially between about 12 and 16 miles (about 20 to 25 km). This is the ozone layer.

Ozone is extremely reactive and, therefore, it is destroyed rapidly at lower levels, where it is more likely to encounter compounds it can oxidize. Above the ozone layer, ozone is destroyed by radiation and below the ozone layer it is destroyed chemically—although it is also formed in the troposphere by photochemical reactions. There is little exchange of gases or particles across the tropopause but, since the 1960s, there have been fears of a gradual accumulation in the stratosphere of substances that could react with ozone. Were this to happen, ozone might be broken down faster than UV radiation causes it to form, and there would be a proportional increase in the amount of short-wave UV radiation reaching the surface. In 1987, evidence of ozone depletion was reported from Antarctica—the so-called "ozone hole."

During the long Antarctic winter, stratospheric air moves in an approximately circular path, with a region of very still air—a vortex—at the center. The temperature in the vortex, about $-130\ °F\ (-90\ °C)$, is low enough to cause water vapor—probably produced by the oxidation of methane—to form polar strato-spheric clouds made from ice crystals. Various nitrogen compounds adhere to surfaces of the ice crystals and, as the winter draws to an end and the returning sun supplies a little energy, these compounds react with free chlorine. This, in turn, forms a stable oxide with ozone, depleting the ozone layer. As spring advances, the air circulation changes, breaking down the vortex. Air containing ozone is drawn in and, with the increasing intensity of the sunlight, the formation of ozone is resumed.

Chlorine is the key ingredient in the polar depletion of ozone. More than half of the chlorine that enters the stratosphere arrives in the form of methyl chloride (CH_3Cl). This is released when surface vegetation is burned—in bush and forest fires, for example—and by many wood rotting fungi. It is chemically stable but, once in the stratosphere, UV radiation breaks the bond linking the methyl (CH_3) and chlorine (Cl) ions, liberating free chlorine. The chlorine from methyl chloride provides a natural background that changes little over the years. The amount of stratospheric chlorine has increased mainly because of the use of chlorofluorocarbon compounds—CFCs in air-conditioning and fire suppression systems. These compounds, of which there are several, are chemically very stable and do not react with the ordinary constituents of the atmosphere. Some CFC molecules are washed to the ground by rain, some enter the oceans, and some adhere to solid particles, such as sand grains, but a proportion remain airborne and seep across the tropopause into the stratosphere.

The ozone layer is thickest in high latitudes in summer. Current predictions, of a possible 10 percent depletion within 50 to 75 years, would increase UV exposure by the equivalent of moving about 100 miles (160 km) toward the equator from high latitudes. There could be a small reduction in yields of sensitive crops and sunbathing could lead to more rapid burning, but the biological consequences are unlikely to be severe. Monitoring suggests, however, that in recent years the intensity of UV radiation over the continents has been decreasing rather than increasing.

2.7   The "Greenhouse" Effect

Long-wave radiation—"black-body" radiation—is absorbed in the atmosphere by a range of gases, but escapes through "radiation windows." This temporary retention of heat by the atmosphere is called the "greenhouse effect," and it serves to keep the earth much warmer than it would be otherwise. If the atmospheric concentration of the "greenhouse gases" increases, however, the existing radiation windows may be partly closed. Provided there was no compensating effect, this would lead to a warming of the atmosphere.

Almost any gas molecules that consist of three or more atoms is likely to be of a size comparable to the wavelength of part of the long-wave electromagnetic spectrum, and such a molecule will absorb some outgoing radiation. Not all molecules are of equal importance. It may be, for example, that a particular gas is already so abundant that it absorbs all radiation in its waveband. Adding more of that gas will have no effect but adding another gas, even in small amounts, may cause radiation to be absorbed in a part of the spectrum that previously was open.

Carbon dioxide is the best known of the greenhouse gases, but this is only because it is the most abundant. It contributes about 50 percent of our present greenhouse effect. Methane contributes about 18 percent, CFCs 14 percent, tropospheric ozone 12 percent, and nitrous oxide 6 percent. Concentrations of all of them are increasing. Over the past century, the proportion of carbon dioxide in the air has increased from about 290 to 350 parts per million. It is still rising by about 1.5 percent a year, mainly because of the burning of carbon-based fuels in facilities and the clearing and burning of forests. Methane, released mainly from cattle and the growing of paddy rice, is increasing at about 1 percent a year. Nitrous oxide is released when organic substances are burned, and ozone is produced by photochemical reactions involving nitrogen oxides and unburned hydrocarbons from boiler flues.

The average global temperature today is about 1 °F (0.5 °C) higher than it was a century ago. This may or may not be due to a greenhouse warming and, beyond predicting a general rise in temperature, calculating the climatic implications of the warming in each region of the world is very difficult. Most scientists agree it would be wise to reduce emissions of greenhouse gases.

3

Basic Building Systems

The building system of a facility is composed of various types of M/E (Mechanical/Electrical) equipment. Systems supporting large installations with closely controlled indoor environments can be very complex. The building system design art has changed in a way that the fine form has not. The architect uses the same basic techniques that have been used for centuries. Meanwhile, the system design engineer has traded in board drafting and hand calculators for a computer and design software. This gives the engineer the capability to analyze libraries of data, perform mathematical calculations, and simulate integrated systems.

Because of the large number of interrelated factors in a system, there can be many solutions to the same building problem, all of which will satisfy the minimum requirements, yet some will be dull and pedestrian while others will display ingenuity and resourcefulness. No single design is the correct one, and for this very reasons it is not entirely desirable to solve a system problem with a step-by-step technique. However, this technique is a good avenue of approach for the uninitiated who lacks the experience necessary to view an entire solution.

The challenge is to involve, but not enslave, the users in the management of their environment. Integration of functions is the answer. In a properly designed system, M/E equipment operates in an integrated manner, so that the output of individual devices will not interfere with or be counter to the performance of the others. Counterproductive performance results in incorrect indoor conditions and wasted energy. Also, maintaining air, power, and water quality is part of internal basic building system design tasks. For example, the protection of sensitive electronic equipment from transients and other interference caused by M/E equipment operation is part of the building system design task.

3.1   The Lighting System

Almost everything scientists know about the universe comes from observing and analyzing light. Yet little was known about light itself until fairly recently. In the seventeenth century, two compelling theories about the nature of light emerged. The corpuscular theory, championed by Sir Isaac Newton, held that light consisted of tiny particles called corpuscles. Another theory proposed that light was a wave, moving through space in much the same way that ripples move across a pond. Although nearly every discovery about light's behavior during the next 200 years seemed to support the wave theory and discredit the corpuscular theory, the advent of quantum physics in the twentieth century reconciled the two theories: Depending on how it is measured and observed, light may assume the characteristics of either a particle or a wave. See Table 3.1 for all light sources and their light intensity.

Light has five distinct properties: propagation, reflection, refraction, diffraction, and interference. Propagation refers to the transmission of light in straight lines. Reflection causes light to bounce off polished surfaces such as mirrors. Light refracts, or bends, when it travels from one substance to another, for example from air through a glass lens. Light waves also will bend around an obstacle's edges, a phenomenon known as diffraction. In interference, intersecting light waves alter each other as they meet. Taken together, these properties explain the workings of devices as varied as high-intensity discharge lighting commonly applied in buildings, lasers, and holograms.

How much light does it take for people to see? We know from experience that it takes more light to see small objects and tasks of low contrast than larger, higher-contrast tasks. As our eyes age, they require more light. In fact, a 50-year-old gets about half as much light on the back of the eye (retina) as a 20-year-old. How much light is required to illuminate work spaces or to perform tasks? The Illuminating Engineering Society of North America (IESNA) provides light level (illuminance) recommendations that are consensus standards agreed up on by lighting professionals.

How does a light bulb glow? The electricity that illuminates lamps—and that runs televisions and appliances—consists of flowing electrons, or current electricity. When a free electron is triggered to move, it occasionally will bump against an atom, exciting the atom, which means it gives some of its energy to the atom. The atom then releases this extra energy as electromagnetic radiation and propels other electrons into action. As electrons flow through the metal filament of an incandescent light bulb, the heating of the filament causes it to give off electromagnetic radiation and glow white hot.

In fluorescent lamps, a current flows through a gas instead of a filament. As the current travels through the gas tube, it causes the gas to give off ultraviolet energy, which excites the phosphor coating inside the tube, triggering a chain reaction that releases electromagnetic radiation as visible light.

Table 3.1   Luminances for Various Light Sources

Light Source	Approximate Average Luminance (cd/m^2)
Natural light sources:	
Sun (at its surface)	2.3×10^9
Sun (as observed from the earth's surface at the meridian)	1.6×10^9
Sun (as observed from the earth's surface at a horizon)	6×10^4
Moon (as observed from the earth's surface at bright spots)	2.5×10^3
Clear sky	8×10^3
Overcast sky	2×10^3
Lightning flash	8×10^{10}
Combustion sources:	
Candle flame (sperm)	1×10^4
Kerosene flame (flat work)	1.2×10^4
Illuminating gas flame	4×10^3
Welsbach mantle	6.2×10^4
Acetylene flame	1.1×10^5
Incandescent lamps:	
A carbon filament (3.15 lm/w)	5.2×10^5
Tantalum filaments (6.3 lm/w)	7×10^5
Tantalum filaments (vacuum lamp 10 lm/w)	2×10^6
Tantalum filaments (gas filled lamp 20 lm/w)	1.2×10^7
Tantalum filaments (projection lamp 26 lm/w)	2.4×10^7
RF (radiofrequency) lamp	6.2×10^7
Black body at 6500 K	3×10^9
Black body at 4000 K	2.5×10^8
Black body at 2042 K	6×10^5
60-W inside frosted bulb	1.2×10^5
10-W inside frosted bulb	2×10^4
Tungsten–halogen sources:	
3000 K CCT	1.3×10^7
3200 K CCT	2.3×10^7
3400 K CCT	3.9×10^7
Fluorescent sources:	
T-8 Bulb 265 mA	1.1×10^4
T-12 Bulb 430 mA	8.2×10^3
T12 Bulb 800 mA	1.1×10^4
T-12 Bulb 1500 mA	1.7×10^4
T-17 Grooved bulb 1500 mA	1.5×10^4
Electroluminescent sources:	
Green color at 120 V 60 Hz	27
Green color at 600 V 400 Hz	68

Table 3.1   (continued)

Light Source	Approximate Average Luminance (cd/m^2)
Carbon arc sources:	
Plain Carbon arc	1.5×10^8
High-intensity carbon arc	1.0×10^9
Enclosed electric are sources:	
High-pressure mercury	1.8×10^6
High-intensity short arcs mercury	2.4×10^8
	$(4.3 \times 10^9$ peak)
The xenon short arcs	1.8×10^8
Clear glass neon tube	1.6×10^3
Clear glass blue tube	8×10^2
Fluorescent tubes	
Daylight and white	5×10^9
Green	9.5×10^3
Blue and gold	3×10^3
Pink and coral	2×10^3

(Reprinted with permission of the IESNA, 120 Wall Street, 17th floor, New York, NY 10005. Taken from the *IESNA Lighting Handbook, 8th Edition.*)

Lighting quality is complex because it considers visual aspects that are highly subjective and not easily quantified. We can, however, understand certain quality characteristics. For example, ceiling reflections detract from lighting quality by obscuring task details by reducing contrast. It is sometimes called a reflected glare and is most noticeable from luminaries located in front of and above the viewing task.

Lamp color also affects lighting quality. Recommendations regarding pleasant combinations of lamp color, temperature, and illuminance are changing and are best left to building occupant preferences. However, we do know that when lamps of good color rendering are used, illuminance may be lowered to achieve equivalent brightness and visual clarity. When upgrading from cool white lamps to higher CRI T8 lamps this effect will be noticeable by employees who may respond that the new lamps are "too bright."

The lamp flicker can also reduce lighting quality. The predominant source of a lamp flicker is from fluorescent lamps operating on magnetic ballasts. The lamps turn on and off 120 times a second, producing distraction, eyestrain, and headaches. The flicker is especially noticeable at high light levels, such as industrial inspection lighting. Electronic ballasts that operate fluorescent lamps at high frequency can reduce the flicker to an imperceptible level. To improve lighting quality, it is important to balance office lighting for visual performance and for visual comfort. Too much contrast will cause workers to be restless. Too little contrast (flat) causes a loss of detail and things appear dull. In retail lighting applications, the lighting lacks the power to attract, and sales will be reduced. Balanced lighting is achieved by avoiding over-diffuse or too strongly directional lighting.

3.1.1 Lighting Terminology and Common Definitions

Lumen Output: The lumen is a unit of luminous flux that is a measure of the total light from a source.

Luminous Efficacy: The efficiency of light sources, calculated by dividing the light output (in lumens) by the power input (in watts), commonly called lumens per watt (lm/w). The higher the lm/w, the more efficient the light source.

Average Rated Life: This is the median value of life expectancy assigned to a lamp, in hours, at which half of a large group of lamps have failed. Any particular lamps or group of lamps may vary from the published rated life. For discharge light sources such as fluorescent and HID lamps, the average rated lamp life is affected by the burn cycle, or the number of starts and the duration of the operating cycle each time the lamp is started.

Lumen Depreciation: Also known as lumen maintenance. All light sources used for interior lighting lose their ability to produce light as they age. Lamp lumen depreciation (LLD) is a dimension less decimal value representing the percent of initial lumens and is one of several light loss factors used in lighting calculations. LLD can be calculated by dividing the mean (design) lumens by the initial lumen rating. The initial lumens are measured after a burn-in time of 100 hours. Values of initial and mean (design) lumens may be found in manufacturers' lamp catalogs.

For example, a 32-watt T8 lamp has an initial lumen rating of 3200 lumens (255 candle power) and a design lumen rating of 2800 (223 candle power). LLD = 2800/3200 = 0.875. This means that the T8 lamp will retain 87.5 percent of its initial light output after 40 percent of its average rated life. Lumen depreciation is affected by the ballast used, line-voltage tolerances, and burn cycles.

Color Temperature: The color temperature of a lamp is described in terms of its appearance (when lighted) to the eye; whether it appears "warm" or "cool." It is measured on a Kelvin scale ranging from 1500 K, which appears red-orange, to 9000 K, which appears blue. Light sources lie somewhere between these two, with those of higher color temperature than 4100 K appearing "cool" and those of a lower color temperature than 3100 K being "warm."

Color Rendering Index (CRI): Color renditions describe the effect a light source has on the appearance of colored objects. The color-rendering capability of a lamp is measured as the CRI. The higher the CRI, the less distortion of the object's color by the lamp's light output. The scale ranges from 0 to 100. A CRI of 100 indicates that there is no color shift as compared to a reference source. The lower the CRI, the more pronounced the shift may be. CRI values should only be compared among lamps of similar color temperature.

Foot Candles: The quantity of light that falls on a work surface is measured in foot candles. One foot candle is equal to one lumen per square foot of area. An illuminance meter is a useful tool to measure the amount of light in work spaces. It is important to understand that the foot candle measure indicates only a level of illuminance. It does not measure the amount of energy to produce that level of light and does not measure the quality of the light produced.

Visual Comfort Probability (VCP): A rating of lighting systems that is expressed as a percentage of people who, when viewing from a specified location and in a specified direction, will find the lighting system acceptable in terms of discomfort glare. The minimum recommendation for office interiors is 70, and the recommendation for computer applications is 80.

3.1.2 Lamps

The term "lamp" is used to describe light sources commonly called light bulbs and tubes. The total light from a light source is measured in lumens. Lamps are now labeled with measured lumen ratings and efficiency ratings (efficacy). Incandescent, fluorescent, and high-intensity discharge (HID) lamps are the most common lamps used for building lighting systems.

Incandescent Lamps: The oldest practical lamp type, the incandescent lamp is what most of us grew up within our homes. They are inexpensive and available in hundreds of sizes, shapes, and wattages and are easily dimmed. However, they are very inefficient—as low as 8 lm/w—have short lamp life, and contribute to building heat.

Incandescent lamps are voltage sensitive, with lamp life, lumen output, and wattage dependent on the applied voltage. Those whose buildings have higher than normal voltage or who want to decrease relampings sometimes use higher voltage rated lamps, usually 130 V. If 130 V lamps are used on 120 V circuits, lumen output will be lower and life will be longer.

Tungsten–halogen lamps are more efficient than standard incandescent lamps. A halogen fill gas combines with the tungsten molecules that boil off the filament. The resulting halogen cycle increases the lm/w, produces whiter light, longer life, and lowers the LLD.

Fluorescent Lamps: The fluorescent lamps are the most commonly used lamp type in commercial and industrial applications. Fluorescent lamps are an electric discharge source in which light is generated when ultraviolet (UV) energy from a mercury arc strikes a fluorescent phosphor on the inside surface of the tube. The tube contains mercury vapor at low pressure and a small amount of inert argon gas (or krypton in reduced-wattage lamps). Characteristics of fluorescent lamps are long life (12,000 to 20,000 hours), high efficacy (75 to 90 lm/w), and excellent color rendering, especially with the newer rare-earth (RE) lamps. Fluorescent lamps are, however, temperature-sensitive and their rated life depends on the hours per start. A ballast is required to properly start and operate fluorescent lamps.

Compact Fluorescent Lamps (CFLs): The generic name for a family of single-ended fluorescent lamps of folded or bridged tube design with high CRI and long life (10,000 hours). Originally designed as preheat lamps, with the starter built into the base, the lamps are now available in rapid-start models. They are often used as alternatives to incandescent lighting. Configurations now include twin tubes, quad tubes, and triple tubes in both preheat and rapid-start models.

High-Intensity Discharge Lamps (HID): Lamps are classified as electric arc discharge lamps that operate under high pressure and generate their light directly from the arc. The arc is contained in a small arc tube that is enclosed in a larger outer glass bulb. The outer glass may be clear or coated on the inside with a fluorescent phosphor. Included in this classification are mercury vapor, metal halide, and high-pressure sodium lamps (HID lamps do not operate instantly; they require time to strike and, when power is removed, longer to cool down and restrike). The buildings generally require a backup lighting system for HID lighting for public safety. Advantages include high lumen ratings and long life.

The mercury vapor (MV) lamp is the oldest HID source and is now considered obsolete. The disadvantages of MV lamps include a poor color rendition, lumen depreciation, and high mercury content.

Metal halide (MH) lamps are now considered the replacement lamp for MV lamps, which they closely resemble. In the metal halide lamp, the arc tube contains, in addition to smaller amounts of mercury, additives called metal halides that provide a brighter, whiter light by improving both lumen and color performance. It is important for safety reasons to prevent MH lamps from reaching "nonpassive failure" by turning these systems off at least 15 minutes every week, group relamping before the end of rated life, and operating them in the correct position and on matching ballasts. Recent advances in MH lamps provide better color consistency. Efficacy ratings are in excess of 100 lm/w and these lamps have long restrike times, up to 15 minutes. Applications for MH lamps include commercial lighting interiors, especially high-ceiling applications; sports lighting; and building facades.

High-pressure sodium (HPS) lamps are the primary source for industrial lighting, highway, and street lighting. HPS lamps have a characteristic yellow color, high efficacy—60 to 140 lm/w depending on wattage—and a long life rating of 24,000 hours. Standard HPS lamps cycle at the end of their life, indicating the need for replacement. Recent advances in HPS lamps include higher CRI models and models that do not cycle at an end of life. A double arc-tube HPS lamp is available for safety and security applications.

3.1.3 Ballasts

Ballast is required to start and operate discharge lamps, fluorescent and HID. All fluorescent ballasts perform two functions. They provide the right voltage to start the arc discharge, and they regulate the lamp current to stabilize light output. In rapid-start ballasts, a third function is to provide the energy to heat the electrodes. Fluorescent ballasts are provided to operate fluorescent lamps in the following ways:

Preheat: Lamp electrodes are heated prior to the application of a high starting voltage that initiates the arc discharge. Preheat operation is characterized by lamp "flickering" when starting. The starting electrode voltage is applied through a "starter," a thermal switch that, when it opens, applies the high starting voltage across the electrodes. No power is applied to heat the cathodes during operation. Lamps of less than 30 watts are usually operated this way.

Rapid-Start: Lamp electrodes are heated prior to and during operation. The ballast has two windings to provide the proper low voltage to the electrodes during starting and operation. Rapid-start operation is characterized by smooth starting and long lamp life.

Instant-Start: Lamp electrodes are not heated. Instant-start ballasts provide a high open-circuit voltage across the unheated electrodes to initiate the arc discharge. Instant-start operation is more efficient than rapid-start, but as in preheat, lamp life is shorter. Eight-foot "slimline" lamps are operated in the instant-start mode.

The ballasting function has traditionally been reliably accomplished with electromagnetic units that consist of a core of magnetic steel laminations surrounded by two copper or aluminum coils. All ballasts have electrical and magnetic losses. These internal losses and the ability of the ballast to operate the lamps efficiently contribute to the overall efficiency of a ballast. The efficiency of magnetic ballasts has been improved by using low-loss magnetic material and copper windings, resulting in lower internal losses.

Ballast efficiency is regulated by the federal appliance standard that prohibits the manufacture of ballasts that do not meet a minimum ballast efficacy factor (BEF). BEF for a specific lamp/ballast combination is calculated by dividing the percent rated light output by the measured input power in watts. The percent rated light output is found by multiplying the ballast factor by 100 percent. BEF can be confusing because it is meant to compare the performance of ballasts on a specific lamp, and is not of particular value in evaluating efficiency opportunities.

Ballast factor (BF) is the ratio of the lamp lumen output on a commercial ballast to the lamp's rated light output. Fluorescent lamps are rated in lab tests using a loss less, perfect reactor ballast, so ballast factor is needed to derate the catalog rating of lamps.

The efficiency of fluorescent ballasts can be improved beyond the use of energy-efficient magnetic ballasts. Concerns for lighting efficiency have prompted the introduction of electronic ballasts. When fluorescent lamps are operated by an electronic ballast at high frequency, they convert the input power to light output more efficiently. The lm/w of the lamp/electronic ballast combination is increased, which means either producing more light for the same power, or producing the same light with lower power. Generally, the lower the ballast factor, the lower the input watts. Proper selections of all electronic ballast parameters are important for successful lighting applications. The total system input watts can actually be less than the total of the lamp wattage. In fact, when a 32-watt T8 lamp is operated at high frequency, it consumes only about 28 watts, making the system input about 30 watts. The actual watt input is controlled by the ballast factor, which can range from 47 to 130 percent.

3.1.4   Luminaires

Unlike fixtures, luminaires are complete lighting units consisting of lamps and ballast to convert power to light, lamp holders, an optical system, a means for connecting to power, and a housing. Each luminaire manufacturer provides light distribution data on the integrated assembly called luminaire photometric.

The *photometric* data for a luminaire provides the light distribution in the form of a polar graph and a table whose values represent the variation in candlepower of a luminaire in a given reference plane. This light distribution is the heart of a photometric report generated from laboratory testing by the luminaire manufacturer or an independent testing laboratory. The report also provides luminaire efficiency, zonal lumen output, a coefficient of utilization (CU) tables, and spacing criteria (SC).

Luminaire Efficiency: Defined as the percentage of lamp lumens that leave the luminaire. This rating quantifies the optical and thermal effects that occur within the luminaire under standard test conditions. Judging luminaires on efficiency alone is a faulty ideal, since the most efficient luminaire is a bare lamp.

Coefficient of Utilization (CU): The percentage of lamp lumens that are received on the work plane, and is a function of luminaire efficiency, room geometry, and room surface reflectance. CU values are used to evaluate how effectively a luminaire delivers light to the work plane in a given space. It is inappropriate to judge a luminaire only by its CU, since the CUs are a function not only of efficiency, but also of its application.

Spacing Criteria (SC): Provides the information regarding how far apart luminaires may be spaced to maintain uniform lighting. To use the spacing criteria, multiply the net mounting height by the spacing criterion's value. The resulting number represents the maximum center-to-center distance that the luminaires may be spaced.

The Luminaire Optical System: Includes the lamp cavity and diffusing media and includes one or more of the following components: reflectors, refractors, lenses, baffles, or louvers. Reflectors redirect light by using the principle of reflection. A refractor is a component that redirects light by refraction. Lenses are made of a transparent or translucent material that has a prismatic configuration on the bottom surface to alter the directional characteristics of light passing through it. Polystyrene lenses are the least expensive, but yellow owing to the ultraviolet (UV) radiation from lamps. Lenses made from acrylic are light stabilized and do not discolor. Translucent sheets of milky white plastic, called diffusers, scatter light uniformly in all directions below the ceiling plane, reducing source brightness and shielding the lamps. And baffles are one-way louvers made of metal or plastic.

Louvers: Consist of baffles, arranged in a geometric pattern, that control the light distribution and shield the source from view at certain angles. Products range from small-cell plastic louvers to large-cell aluminum louvers.

Parabolic luminaires use large-cell louvers in a parabolic shape. The resulting light distribution reduces the glare, controls light output, and produces high aesthetic appeal. Paracube louvers are small metallic-coated plastic squares; one-inch squares are the most common size. Small-cell louvers have high visual comfort probability (VCP), but sacrifice luminaire efficiency and spacing criteria. The highly polished surfaces used in some louvers and reflectors are called specular. Semispecular finishes reflect some of the light directionally, with some amount of diffusion.

Luminaires are also classified according to the manner in which they control or distribute their light output. They can be direct (downward), indirect (upward) or direct–indirect.

Direct luminaires can be open or shielded. Shielded luminaires use lenses, baffles, or louvers. Down lights are direct ceiling luminaires that direct all the light downward. They can be recessed, surface-mounted, or suspended. Recessed units include HID down lights and a wide range of fluorescent troffers. Recessed troffers are luminaires mounted within the ceiling so that the bottom surface is in the same plane as the ceiling. Some troffers are designed to lie on the tee-bar flanges of a suspended-grid ceiling system. Surface-mounted troffers are attached directly on the ceiling and project below the ceiling by their depth. Pendant-mounted units, such as direct–indirect, are supported from the ceiling structure by stems or pendants that position the luminaire below the ceiling. Air slot is a term applied to some recessed luminaires that supply or return comfort air. This is accomplished through slots. A static luminaire does not handle comfort air.

Indirect luminaires radiate light up to a reflecting ceiling. Indirect lighting systems usually use luminaires suspended from the ceiling by aircraft cables. Cove lighting, wall-mounted units, and furniture-top units are also considered indirect. Lighting offices with indirect luminaires results in higher, more uniform brightness on both ceilings and walls, and computer screens' glare is reduced significantly. Shadows are reduced from partitions and under-shelf cabinets. The success of indirect systems depends on maintaining a high ceiling reflectance. Recent designs use T8 lamps and electronic ballasts in a low-profile housing with lenses and special reflectors to achieve high luminaire efficiency and a broad distribution, while allowing mounting close to the ceiling.

Direct–indirect luminaires combine the efficiency and high coefficient of utilization of direct luminaires with the uniformity and glare control of indirect luminaires. Some office and school systems are designed for an equal balance of direct down light and indirect up light.

Luminaire Light-Loss Factors (LLFs): The light lost due to reflecting surfaces degrade over time. This degradation of the surface is nonrecoverable. In contrast, luminaire dirt depreciation (LDD) is the recoverable light loss factor that describes how light is lost from the initial illuminance provided by clean, new luminaires compared with the reduced illuminance that will be provided after dirt collects on the reflecting surfaces at the time when it is expected that cleaning will be done.

3.1.5 Lighting and Energy

To understand the contribution of lighting to building cooling loads it is important to know that three main energy conversions occur during the process of generating light by a fluorescent lamp. Initially electrical energy is converted into kinetic energy, exciting UV radiation. This UV energy in turn is converted to visible energy by the lamp phosphor. During each of these conversions some energy is lost so that only small percentage of input is converted into visible radiation. Figure 3.1 shows the approximate distribution of energy in a cool white fluorescent lamp.

Buildings waste lighting energy because of hearsay beliefs like: "Continuously operating fluorescent lights is cheaper than turning them off for brief periods." Or "Lights shouldn't be turned off because it shortens lamp life and increases maintenance costs." Actually, turning off fluorescent lights saves energy, extends overall lamp life, and reduces replacement costs. Although the average rated life of fluorescent lamps is shortened by switching, calendar life is lengthened. Calendar life is the time between lamp changes and includes the time the lamp is off. For example, standard T32 rapid-start lamps operated continuously result in a rated lamp life of 32,000 hours (calendar life of 3.7 years). Turning off T32 lamps for 12 hours each day decreases the average rated lamp life to 29,000 hours, but calendar life is extended to 6.6 years. The belief that there is a high inrush current during the starting is unfounded and is not even of any theoretical concern if an electronic ballast is used.

3.2 Power Systems

Many factors should be considered in the design of electric power distribution system for a modern commercial/industrial building. Factors that influence the system configuration are characteristics of the load, quality of service available and required, the size of the building, and cost. Basic arrangements used for dis-

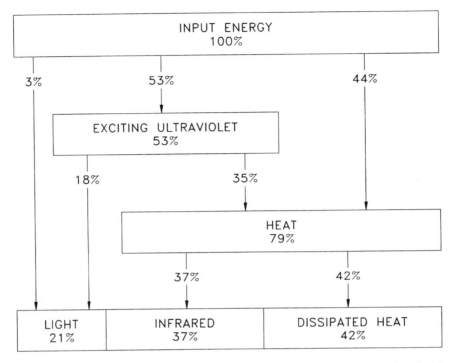

Figure 3.1 Energy distribution in a typical cool white fluorescent lamp. (Reprinted with permission of the IESNA, 120 Wall Street, 17th floor, New York, NY 10005. Taken from the *IESNA Lighting Handbook, 8th Edition.*)

tribution of electric power for commercial/industrial facilities are tabulated in table 3.2. First investigate power distribution requirements of the building and then select the best system or combination of systems based on the needs of the building function. The commercial and industrial power system classification for voltage classes are: Medium voltage is 2400 volt up to and including 69,000 volt. High voltage refers to 115,000 up to 230,000 volts. Unfortunately, the common usage of the term "high voltage" by the average user can be anything more than 120 volts or anything more than 600 volts. The discussion on system arrangements described in table 3.2 covers both medium voltage and low voltage distribution.

Table 3.3 lists the US standard nominal voltages in common use in the buildings. The numbers listed in the right-hand column are used in equipment ratings, but these should not be confused with numbers designating the nominal system voltage on which the equipment is designed to operate.

Table 3.2 Types of Power Distribution System

Power System Type	Features	Discussion
Simple radial system (Figure 3.2)	Distribution is at the utilization voltage. A single primary service and distribution transformer supply all the feeders. There is no duplication of equipment. System investment is the lowest of all circuit arrangements. Operation and expansion are simple. When quality components and appropriate ratings are used reliability is high. Loss of a cable, primary supply, or transformer will cut off service. Equipment must be shut down to perform routine maintenance and servicing.	This system is satisfactory for small industrial installations where process allows sufficient down time for adequate maintenance and the plant can be supplied by a single transformer.
Expanded radial system (Figure 3.3)	The advantages of the simple radial system may be applied to larger loads by using an expanded radial primary distribution system to supply a number of unit substations located near the load, which in turn supply the load through radial secondary systems.	The advantages and disadvantages are the same as those described for the simple radial system.
Primary selective system (Figure 3.4)	Protection against loss of a primary supply can be gained through use of a primary selective system. Each unit substation is connected to two separate primary feeders through switching equipment to provide a normal and an alternate source. On failure of the normal source, the distribution transformer is switched to the alternate source. Switching can be either manual or automatic, but there will be an interruption until the load is transferred to the alternate source.	Cost is somewhat higher than that of a radial system because of duplication of primary cable and switchgear.

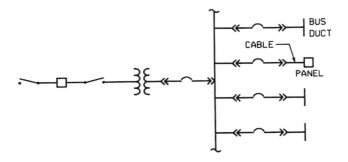

Figure 3.2　Simple radial system. (Copyright © 1994. IEEE. All Rights Reserved).

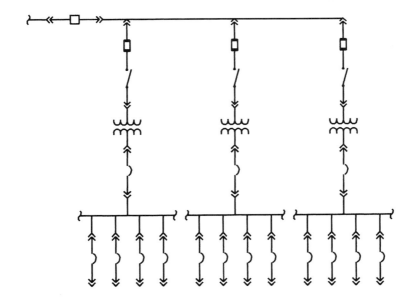

Figure 3.3　Expanded radial system. (Copyright © 1994. IEEE. All Rights Reserved.)

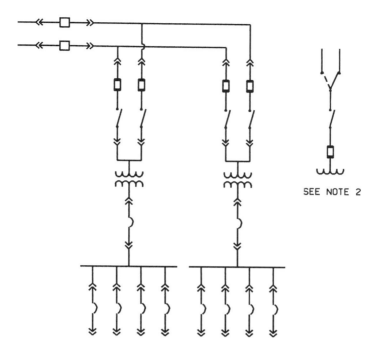

SEE NOTE 2

NOTES:
1. IF NON-DRAW-OUT FUSED SWITCHES ARE USED,
   THE FUSE SHOULD BE ON THE LOAD SIDE OF THE SWITCH.
2. AN ALTERNATE ARRANGEMENT USES A PRIMARY SELECTOR
   SWITCH WITH A SINGLE FUSED INTERRUPTER SWITCH
   (WHICH MAY NOT HAVE CERTIFIED CURRENT SWITCHING ABILITY).

Figure 3.4   Primary selective system. (Copyright © 1994. IEEE. All Rights Reserved)

Table 3.2   (*continued*)

Power System Type	Features	Discussion
Primary selective system (*continued*)	If the two sources can be parallel during switching, some maintenance of primary cable and switching equipment, in certain configurations, may be performed with little or no interruption of service.	

Table 3.2 (*continued*)

Power System Type	Features	Discussion
Primary loop system (Figure 3.5)	A primary loop system offers improved reliability and service continuity in comparison to a radial system. In typical loop systems, power is supplied continuously from two sources at the ends of the loop. Such a system, if properly designed and operated, can quickly recover from a single cable fault with no continuous loss of power to utilization equipment. Single electrical power can flow in both directions in a loop system. It is essential that detailed operating instructions be prepared and followed. Additionally, if the two supply points for the loop originate from different buses, the design must consider available short-circuit capacity from both buses, the ability of both buses to supply the total load, and the possibility of a flow of current from one bus to the other bus over the loop.	To realize optimum service reliability of a primary loop system, the system should be operated (closed-loop mode). A cable fault within the loop may be automatically isolated without loss of transformer capacity. No loss of power will occur, although the system will experience a voltage dip until the circuit breakers clear the fault. A primary loop system may be operated with one of the series switches in open. A disadvantage of open-loop operation is that a cable failure will result in the temporary loss of service to some portion of the system.
Secondary selective system (Figure 3.6)	If pairs of substations are connected through a secondary tie circuit breaker, the result is a secondary selective system. If the primary feeder or transformer fails, supply is maintained through the secondary tie circuit breaker. The tie circuit breaker can be operated in a normally opened or a normally closed position. In case of the normally opened tie circuit breaker, voltage is maintained to the unaffected transformer's circuits. In the case	Normally the systems operate as radial systems. Maintenance of primary feeders, transformer, and main secondary disconnecting means is possible with only momentary power interruption, or no interruption if the stations can be operated in parallel during switching, although complete station maintenance will require a shutdown. With the loss of one primary circuit or transformer, the total sub-

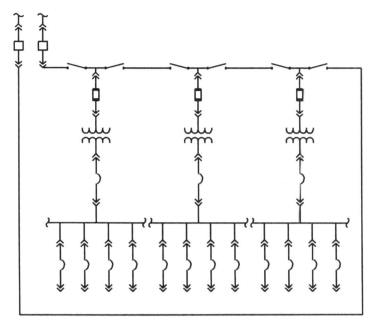

Figure 3.5 Primary loop system. (Copyright © 1994. IEEE. All Rights Reserved)

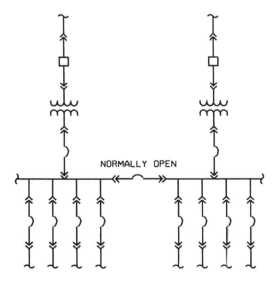

Figure 3.6 Secondary selective system. (Copyright © 1994. IEEE. All Rights Reserved)

Table 3.2 (*continued*)

Power System Type	Features	Discussion
Secondary selective system (*continued*)	of the normally closed tie, a voltage depression occurs on the bus until the affected transformer's circuit breaker opens.	station load may be supplied by one transformer. The secondary selective system may be combined with the primary selective system to provide a high degree of reliability. This reliability is purchased with additional investment and addition of some operating complexity.
Secondary spot network (Figure 3.7)	In this system two or more distribution transformers are each supplied from a separate primary distribution feeder. The secondaries of the transformers are connected in parallel through a special type of device, called a network protector, to a secondary bus. Radial secondary feeders are tapped from the secondary bus to supply utilization equipment. If a primary feeder fails, or a fault occurs on a primary feeder or distribution transformer, the other transformers start to feed back through the network protector on the faulted circuit.	The secondary spot network is the most reliable power supply for large loads. A power interruption can occur only when there is a simultaneous failure of all primary feeders or when a fault occurs on the secondary bus. There are not momentary interruptions caused by the operation of the transformer switches that occur on primary selective, secondary selective, or loop systems. Voltage sags caused by large transient loads are substantially reduced. Networks are expensive because of the extra cost of the network protector and duplication of transformers capacity. In addition, each transformer connected in parallel increases the short-circuit-current capacity and may increase the duty ratings of the secondary equipment. This scheme is used only in low-voltage applications with a very high load density.

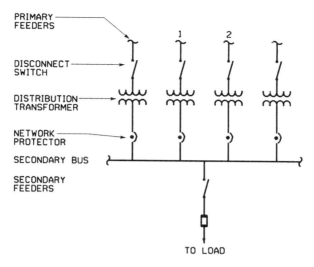

Figure 3.7   Secondary spot network. (Copyright © 1994. IEEE. All Rights Reserved)

Table 3.2   (*continued*)

Power System Type	Features	Discussion
Ring bus (Figure 3.8)	The ring bus offers the advantage of automatically isolating a fault and restoring service. A fault anywhere in the ring results in two interrupting devices opening to isolate the fault.	The ring bus scheme is often considered where there are two or more medium voltage (i.e., 4.16, 4.8, or 13.2/13.8 kV) distribution services to the facility and the utmost in flexibility and switching options are desired. Manual isolating switches are installed on each side of the automatic device. This allows maintenance to be performed safely and without interruption of service. This will also allow the system to be expanded without interruption.

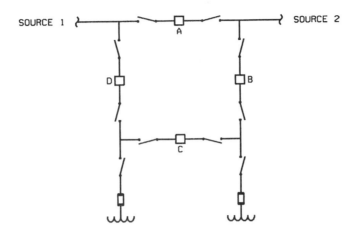

Figure 3.8   Ring bus system. (Copyright © 1994. IEEE. All Rights Reserved)

3.3   Heating, Ventilation, and Air Conditioning (HVAC) Systems

HVAC systems maintain desired environmental conditions inside buildings. In almost every application, there are several ways these conditions may be maintained. Ideally HVAC systems permit people or products to function within the buildings at optimum level. Different systems approach this ideal with varying degree of success. HVAC systems are categorized and segregated by equipment arrangement and how they control environment in the space. Water systems use hot or chilled water to convey heat to or from a controlled space through piping system, connecting a source such as boiler, water heater, or chiller, to a suitable terminal heat transfer units. Hot water heating and chilled water cooling systems are frequently called hydronic systems. Water systems are classified by temperature, flow generation, pressurization, or piping and pumping arrangement. They are either closed (recirculating) or once through (open) systems. Buildings mostly use recirculating systems.

Dampers and fans are used to control the flow of air into, out of, and within a building. Necessary pressures to move air within ductwork and conditioned spaces are provided by fans, operated either constantly or intermittently. Table 3.4 lists, air handling systems commonly used in commercial buildings. Forced-air systems range in complexity from single duct, single fan networks, found in private residences and small buildings, to sophisticated Air Handling and Variable Air Volume (VAV) systems found in large buildings.

Both CV and VAV systems are important in their own right. Hybrid systems have been developed that combine the advantages of CV and VAV systems and in turn minimize the disadvantages. VAV systems with terminal reheat fans pow-

Table 3.3   Nominal Power System Voltages

Standard Nominal System Voltages	Associated nonstandard Nominal system voltages
Low voltages	
120	110, 115, 125
120/240	110/220, 115/230, 125/250
208Y/120	216Y/125
240/120	
240	230, 250
480Y/277	460Y/265
480	440
600	550, 575
Medium voltages	
2400	2200, 2300
4160Y/2400	
4160	4000
4800	4600
6900	6600, 7200
8320Y/4800	11,000, 11,500
12,000Y/6930	
12,470Y/7200	
13,200Y/7620	
13,200	
13,800Y/7970	14,400
13,800	
20,780Y/12,000	
22,860Y/13,200	
23,000	
24,940Y/14,400	
34,500Y/19,920	
34,500	33,000
46,000	44,000
69,000	66,000
High voltages	
115,000	110,000, 120,000
138,000	132,000
161,000	154,000
230,000	220,000

Table 3.4 Brief Tabulation of Variety of Air Handling Systems Used in Buildings

System Type	Features	Comments
Single zone system (Figure 3.9)	AHU serving a single air-conditioned space	Unable to satisfy different needs in different spaces at any given time. Does not heat/cool simultaneously.
Constant volume (CV) terminal reheat system	Total supply air is cooled to satisfy worst condition at any given time.	Wastes both cooling and heating energy but offers comfort in the conditioned spaces.
Mixed air systems (Figures 3.10 and 3.11) Double duct (DD) system Multizone (MZ) system	With MZ system, temperature control for all zones is centrally located at AHU, whereas in DD system, the space temperature control is in the mixing terminals located over the occupied spaces.	Offers a good temperature control but limited humidity control. Addition of a bypass plenum to hot and cold plenums reduces thermal energy waste.
Induction systems (Figure 3.12)	In the induction unit primary air from central AHU is discharged through an induction nozzle at high velocity, which induces a certain amount of room air and supplies the mixture into the conditioned space.	Uses more air transportation energy (fan HP).
VAV systems (Figures 3.13 and 3.14)	Supply air terminals in a VAV system satisfy space cooling needs by modulating the amount of volume of air rather than varying temperature of the supply air. Since building skin losses are continuous linear functions of the outside air temperature, the use of a CV perimeter heating/cooling system is very common.	A VAV fan HP requirement is considerably smaller than a CV system. Variable volume, variable temperature (VVVT) is suitable for smaller buildings in mild climate areas. Using a VVVT system for large buildings or in extreme climates may cause control problems.

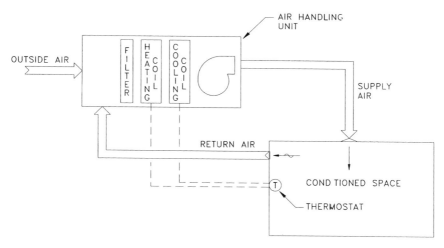

Figure 3.9   Single-zone air handling system.

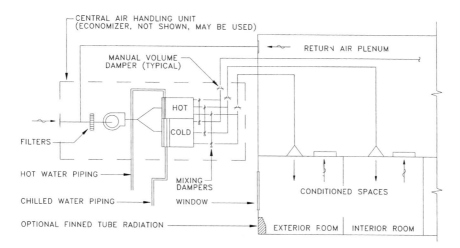

Figure 3.10(a)   Multizone system with mixing dampers (only three zones shown for clarity).

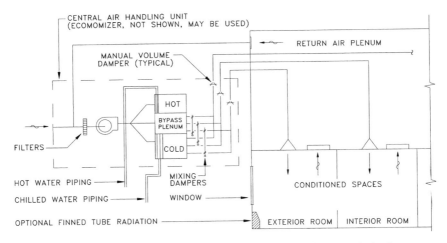

Figure 3.10(b) Triple deck multizone system with mixing dampers (only three zones shown for clarity).

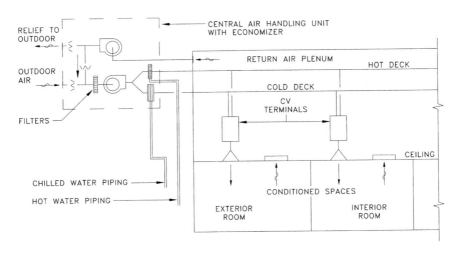

Figure 3.11(a) Constant volume single-fan double duct system.

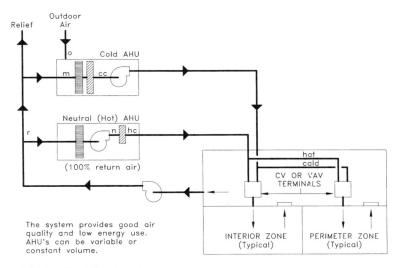

Figure 3.11(b)   Dual-fan dual-duct system.

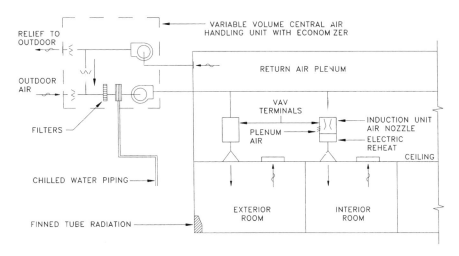

Figure 3.12   Variable volume system with independent perimeter heating.

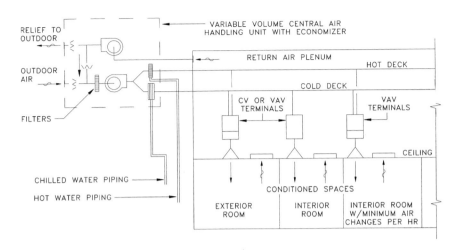

Figure 3.13   Variable volume double duct system.

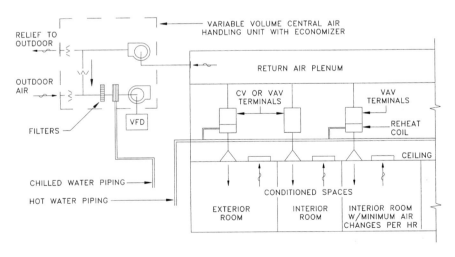

Figure 3.14   Variable volume terminal reheat system.

ered boxes are a good example of a hybrid system. Thermal and transportation savings in these systems are achieved during the period when the system is operating in the VAV mode, that is, cooling only.

See Table 3.5 for recommended (or frequently used) air handling systems in various types of buildings with varied functions.

The following list provides basic configurations of air, water and steam HVAC systems typically used in buildings:

All-Air Systems

    Single-path systems

        Single-duct, variable air volume

        Single-duct, variable air volume—induction

        Single-duct, variable air volume—fan powered

    Dual-Path Systems

        Dual duct

        Multizone systems

Air-and-Water Systems

    Air–water induction systems

    Air–water fan-coil systems

    Air–water radiant panel systems

All-Water Systems

    Two-pipe systems

    Three-pipe systems

    Four-pipe systems

Unitary Refrigerant-Based Systems

    Window-mounted air conditioners and heat pumps

    Through-the-wall mounted air conditioners and heat pumps

    Air-to-air heat pumps

    Outdoor unitary equipment systems

    Indoor unitary equipment systems

    Water-source heat pump systems

Heat Recovery Systems

    Heat pumps and heat recovery chillers

Energy recovery wheels
Runaround heat recovery loops

Panel Heating and Cooling Systems
Radiant heating panels
Perimeter heating system
Embedded pipe coil system
Ceiling's electric resistance panel
Forced warm air floor panels

Congeneration Systems
Prime movers
Reciprocating internal combustion engines
Combustion gas turbines
Expansion turbines
Steam boiler turbine combinations
Generators
Waste heat recovery systems
Electrical and thermal distribution system

Steam Systems
Convection-type steam-heating systems
Combined steam and water systems
One-pipe steam heating systems
Two-pipe steam heating systems

Central Plant Heating and Cooling System
Central chilling plant
Chilled water and dual-temperature systems
Terminal heating and cooling units

Chilled and Dual-Temperature Water Systems
Two-pipe chilled water systems
Brine systems
Two-pipe dual-temperature water systems

Two-pipe natural cooling systems
Three-pipe systems
Four-pipe systems
Condenser water systems
Once-through systems
Cooling tower systems

Medium- and High-Temperature Water Heating Systems
Direct fired hot water generators
Direct contact heating

Infrared Radiant Heating System
Spot heating system
Total building heating system

Geothermal Heating/Cooling System
Ground source heat pump system
Ground coupled heat pump system
Horizontal system
Parallel system

Thermal Storage System
Ice storage system
Chilled or hot water storage system
Building mass thermal storage system
Electrically charged heat storage system

Solar Energy Utilization
Solar heating and cooling systems
Solar collector and storage subsystems
Solar cooling system

Energy Management System

Special Ventilation System

Automatic Control System

Sound and Vibration Control System

Water Treatment System Evaporative Air Cooling System
　Indirect/direct evaporative cooling systems

Snow Melting System
　Embedded hydronic pipe system
　Embedded electric resistance heating
　Overhead high-intensity infrared radiant heating

Smoke Management System
　Stack effect system
　Compartment system
　Smoke dilution system
　Pressurization system

3.4　Security Systems

Security Systems (SS) Access Control: There are numerous methods of providing access control. Among these methods are guard force, mechanical or electronic cipher locks, dial combination locks, key locks, and internally controlled electrical door releases. Recent developments in biometric access control can be used for access when permitted by individual service publications. Biometric access control is the method of identification and verification in which the person seeking access is identified by fingerprints, palm pattern and geometry, retinal pattern, voice analysis, signature dynamics, and other methods.

　Closed Circuit Television (CCTV): This type of system may be used for physical security inside and outside the facility. This system is particularly effective for monitoring hallways, doors, and perimeters. The CCTV system consists of a monitor station and cameras installed in strategic locations.

　Intrusion Detection System (IDS): An intrusion detection system may be used in facilities to provide additional protection. IDSs typically consist of magnetic switches and motion sensors. All exterior entrances, open areas, and sensitive areas within the facility should be protected. Motion sensors may be installed to provide general coverage of open areas to prevent an intruder from crawling under the area of coverage. The IDS should connect to a remote monitor station through a dedicated line or autodial telephone line. This will aid in providing notification to the appropriate police or security agency.

Table 3.5 Heating, ventilating, and air-conditioning practices for a broad range of applications

Application	Single Zone Self-Contained Rooftop[a]	Through-Wall	Multizone[b]	Constant Volume Reheat	Variable Air Volume Single Duct[c] W/Separate Perimeter Heat[d]	Fan-Powered Parallel[e]	Fan-Powered Series[f]	Dual Duct Constant Volume w/ Reheat[g]	Two-Pipe Fan Coil, Separate O.A.[h]	Four-Pipe Fan Coil Separate O.A.[h]	Water Source Heat Pump Central	Modular Distribution[b]
Specialty Stores	+				+	o	o					+
Restaurants	+				+	o	o			+		
Bowling alleys	+				o	+	+					
Radio/TV studios	+		o	+								
County clubs	+		+		+							
Funeral homes	+		+		o							
Churches	+		+		+							
Theaters	+				+					o	+	
Office buildings	o[i]		o[i]		+	+	+			+	+	+
Hotels, dormitories	o	o										
Motels/apartment buildings		+							+	+		+
Hospitals	o			+[j]	+[l]	+[l]	+[l]	+[j]	o	+[k]		
Schools and colleges	o		o	+	+	+				+	o	
Museums	o		+	+	o	o				+	o	+
Libraries, standard	+		+		+	+	+					
Libraries, rare books	o		+		o	o	o					
Department stores	+		o		+	+	+					
Laboratories	+			+	+	+	+				o	o

+ = Systems frequently used; o = System occasionally used

[a] In cold climates, often in conjunction with separate perimeter heating system due to high window/wall heat losses and/or to avoid central fan operation during extended unoccupied periods.

[b] Not recommended for use in hot/humid climates (poor RH control at high cooling load/high outside air humidity level). Limit 12 zones per unit.

[c] Pressure-independent boxes recommended for most applications involving long duration occupancy (fewer expensive pressure-dependent boxes are typically used in short-duration occupancies such as stores and supermarkets, where drafts are tolerable).

[d] Cooling-only VAV box typically used for large interior zones or in perimeter zones provided with separate heating system.

[e] Fan powered parallel VAV boxes (typically used in perimeter zones) are generally more energy efficient than series type VAV boxes. Since the small box fans operate only during the heating cycle when fan motor losses contribute to the heating function.

[f] Fan-powered series VAV boxes (typically used only for perimeter zones) should be used only where constant air flow to the occupied space is very important design criterion (small, inefficient fan motors operate continuously even during the cooling cycle, increasing annual fan power consumption).

[g] Not recommended in regions where low cooling loads coincide with high outside air humidity (RH levels in space increases quickly). Use in areas with moderately humid climates and less than 78°F (25.5 °C) outside air dry bulb temperature at design condition.

[h] Modular fan coil units and water-source heat pumps require outside air ventilation in all but residential applications. Separately ducted ventilation systems offer better control, filtering, energy efficiency, and less maintenance than through-wall outside air louvers.

[i] Appropriate for small, one or two story buildings only.

[j] Appropriate only for ORs, ICU, nurseries, delivery rooms, and similar spaces requiring exceptionally high degree, or in building applications where recovered heat can be used for reheating function.

[k] Used in isolation wards to avoid recirculation of air to other areas.

[l] Appropriate for general areas and patient rooms.

O.A. = outside air

3.5 Basic Fire Protection Systems (FPS) Used in Buildings

It is necessary to install regularly in buildings automatic fire extinguishing systems that have a fire alarm signaling system. Fire alarm signals where included initiate extinguishing agent discharge, alarm building personnel, and also provide supervisory service to indicate any off-normal conditions of the system. There are many types of extinguishing systems used with a fire alarm system. Figure 3.15 lists all types of fire protection systems encountered in buildings. At the top are shown regulations and authorities that classify the building and owner's underwriter requirements that determine the type and need.

Sprinkler systems are used in the majority of commercial, institutional, and industrial buildings. There are four types of automatic water sprinkling systems: (1) wet-pipe, (2) dry-pipe, (3) pre-action, and (4) deluge. Wet-pipe systems employ automatic sprinklers attached to pipes connected to a water supply. The pipes are filled and the sprinklers discharge water immediately when opened by the heat of a fire. The dry-pipe system contains pressurized air or nitrogen. When the sprinklers are opened, as by the heat of a fire, the gas pressure drops water pressure from the supply to open a valve, and water flows into the pipes to be dispensed from the sprinklers. See Figure 3.16 for different types of valves. The pre-action system employs automatic sprinklers connected to pipes containing air that may or may not be under pressure. A supplemental fire detection system is installed in the same area as the sprinkling system. When the detection system is activated, a valve opens, permitting water to flow into the pipes and to be discharged through the sprinklers (which have been opened by the fire). The deluge system is similar to the pre-action system, but uses open sprinklers instead of closed ones.

Table 3.6 provides a brief description of some important fire protection systems listed in the Figure 3.15.

3.5.1   Fire Alarm, Detection, and Signaling Systems

Fire alarm systems are classified according to the type of functions they are expected to perform. Regardless of the type of alarm system, national or local codes have very specific requirements for protective system signaling and source of power supply and backup power supply. Fire detection and alarm system circuit can be supervised or nonsupervised. Table 3.7 summarizes basic classification of fire detection and alarm system per NFPA into six types: local, auxiliary, remote, proprietary, central and emergency.

3.6 M/E Equipment Arrangement

The M/E systems arrangement is a combination of pieces of mechanical and electrical and communication equipment and their interconnection with conduits,

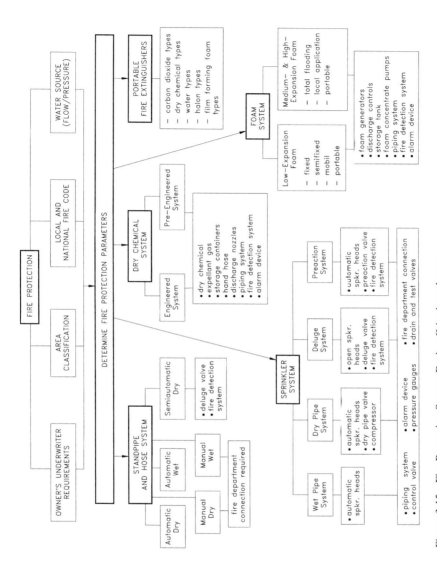

Figure 3.15 Fire Protection System Design Criteria and types

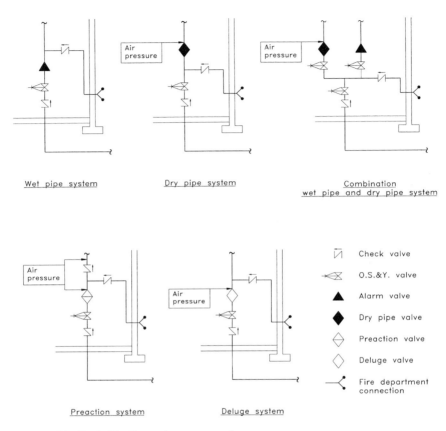

Figure 3.16 Basic Fire Protection system valve arrangements.

pipes, wires, channels, ducts, fibers, and fields. Typically gases, vapors, and liquids are conducted by pipes, tubes, and hoses, and sometimes open channels; electricity is conducted by solid conductors, ionized liquids, and ionized gases; heat and vibration are conducted by all solids; and people are conducted by paths, ladders and steps, elevators, and escalators.

Heat is transported by moving fluids, including furnace gases, boiled water (steam), boiled refrigerants, air (including refrigerated air to be reheated), and the vapor in heat pipes. Some fluid transportation is driven by convection owing to the different densities of warm and cold fluids, and some is driven by pumps and blowers. Vapor is also driven to the cool end by the pressure drop due to condensation.

Information, signals, video, audio, TV, and data communication is increasingly done via light pulses in optical fibers because of the enormously greater

Table 3.6 Basic fire protection systems and applications

Type	Basic Description	Features
1. Wet-pipe sprinkler system	A permanently piped water system, maintained under pressure. When fire occurs, discharges extinguishing medium (water) through heat-activated fusible sprinkler heads exposed to the high heat.	Automatically activates and controls or extinguishes the fire. Protects base structure. Causes water damage to building interiors and equipment and any material or device prone to water damage.
2. Pre-action automatic sprinkler system	A permanently piped system employing heat-activated sprinklers. Piping system is filled with air that may or may not be under pressure. A fire, heat, or smoke detection system installed in the same area of a sprinkler zone. On detection of smoke, fire, or heat it activates a "pre-action valve" that permits water to flow into the sprinkler system piping ready to be discharged from any sprinklers that are opened by the heat from the fire.	Automatically detects fire. Minimizes the accident discharge of water due to mechanical damage to sprinkler heads or piping, and thus is useful for the protection of building interiors and critical equipment.
3. On–off automatic pre-action sprinkler system	A pre-action system with fire/heat detector interlock; heat detector actuates "a pre-action valve" to open at a predetermined temperature and close when normal temperature is restored. Should the fire rekindle after its initial actuation and closing, the valve will reopen and water will again flow from the opened heads. The valve will continue to open and close in accordance with the rate of a temperature rise sensed by the heat detectors.	In addition to the favorable feature of the automatic pre-action wet-pipe system, this has the ability to further reduce the water damage by automatically stopping the flow of water when no longer needed, thus eliminating unnecessary water damage.

Table 3.6 (*continued*)

Type	Basic Description	Features
4. On–off automatic wet sprinkler system	A wet-pipe system with on–off sprinkler heads. Here, each individual head has incorporated in it a temperature-sensitive device that causes the head to open at a predetermined temperature and close automatically when the temperature at the head is restored to normal.	This feature adds the automatic or reclosing control to reduce water damage. It provides the best of wet and pre-action systems.
5. Dry-pipe sprinkler system	It is similar to the pre-action system with no supplemental detection system. It employs "deluge valve," heat-operated sprinklers attached to a piping system containing air under pressure. When a sprinkler operates, the air pressure is reduced, a "deluge valve" is opened by water pressure, and water flows to any opened sprinklers.	Provides protection to areas subject to freezing. Water supply and deluge valves must be in a heated area.
6. Deluge sprinkler system	It employs open sprinklers attached to a piping system that is empty, with a supplement fire detection system installed in the same area as the sprinklers. A deluge control valve is normally closed to prevent water from entering the system. When fire occurs, the supplement detection system actuates and opens the water control valve, allowing water to enter the system. Water is discharged through all of the sprinklers on the system, thus deluging the protected areas.	Used in high hazard areas requiring an immediate application of water over the entire hazard. *Examples:* Flammable liquid handling and storage areas, aircraft hangars, oil refineries, chemical plants, hazardous equipment

Table 3.6 (*continued*)

Type	Basic Description	Features
7. Standpipe and hose system	A permanently installed wet piping systems in a building, maintained under pressure, to which a flexible hose is connected for emergency use by building occupants or by the fire department.	Normally provided as a complement to an automatic wet/dry or pre-action sprinkler system. Building personnel need to be familiar with effective hose application.
8. Non CFC agent flooding system	A permanently piped system using a limited stored supply of a non-CFC inert gas under pressure, and discharge nozzles arranged to totally flood an enclosed area. Released automatically by a suitable heat/fire or smoke detection system. Extinguishes fires by blanketing the fire zone, thus inhibiting the chemical reaction of fuel and oxygen to extinguish fire.	No damage to critical electronic equipment, devices, or other irreplaceable valuable objects. No agent residue. Toxicity level of the agent used governs the limit of agent concentration and duration of extinguishing action. A new breed of non-CFC agents may not extinguish deep-seated fires in ordinary solid combustibles, such as paper, fabrics, etc., but are effective on electrical fires or surface fires in these materials. These systems require special airtight areas/enclosures to maintain concentration for a specified time. The high-velocity discharge from nozzles requires extreme care in location of nozzles to avoid damage due to instant release.

Table 3.6 (*continued*)

Type	Basic Description	Features
9. Carbon dioxide flooding system	Uses carbon dioxide gas as flooding agent. Activates through smoke, fire, or heat detection system. Extinguishes fires by reducing oxygen content of air below combustion chemical reaction points.	Toxic agent, suitable only for normally unoccupied areas and where water damage is more catastrophic, for example, service and utility areas. Personnel must evacuate before agent discharge to avoid suffocation. May not extinguish deep-seated fires, but effective on electrical fires.
10. Dry, chemical flooding system	Similar to other flooding systems, except uses a dry chemical powder. Released by mechanical thermal linkage or detection system. Effective for live electrical fires.	Can suffocate and should not be used in personnel-occupied areas. Leaves powdery deposit on all exposed surfaces. Requires a cleanup. Excellent for service facilities having kitchen range hoods, and ducts. May not extinguish deep-seated fires in ordinary solid combustibles, such as paper, fabrics, etc., but effective on surface fires in these materials.
11. High-expansion foam system	A fixed wet foamy extinguishing system that generates a foam agent through mixing for total flooding of confined spaces. It extinguishes fire by cooling and displacing vapor, heat, and smoke. This reduces the oxygen concentration at the fire.	High volumetric discharge supplemented by high-expansion foam may inundate personnel in the space, blocking vision, hearing, and creating breathing problems. Leaves the residue and requires a heavy cleanup. Effective on oil storage area fires.
12. Water-mist (WM) system	A new type of preactive sprinkler system that uses high pressure to create a small-droplet water mist to extinguish fire.	The small droplets provide cooling, oxygen displacement and radiant heat attenuation. A WM system could be used in computer room fire suppression.

Table 3.7   Basic Fire Alarm Systems and Applications

Category	Basic Description	Features
1. Local boundary alarm system.	Building fire alarm system, operation limited to within building. Alarms on activation of a manual fire alarm box, waterflow in a sprinkler system, tamper switches on fire valves, or detection of a fire by a smoke or heat detecting system.	The main purpose of this system is to provide an evacuation alarm for the occupants of the building. Requires building personnel to transmit the alarm to local fire authorities.
2. Auxiliary fire alarm signaling system	System includes a standard local municipal coded fire alarm box to transmit a fire alarm from building to local municipal fire station. These alarms are received on the same municipal equipment and are carried over the same transmission lines as are used to connect fire alarm boxes located on streets. Operation is initiated by the building fire detection system, manual pull stations, or water flow in the pipes.	Summons immediate help from a local fire department. Some communities are equipped to accept this type cf system and others are not.
3. Remote station fire alarm signaling system	Utilizes building dedicated telephone lines to transmit alarm to a remote station, such as a fire station or a police station. Includes separate receiver for individual functions being monitored, such as fire alarm signal, manual pull station, or sprinkler waterflow alarm.	Requires leased telephone lines into each protected building.

Table 3.7 (*continued*)

Category	Basic Description	Features
4. Central station alarming system	Incorporates a privately owned central station whose function is to monitor the connecting lines constantly and record any indication of fire, supervisory, or other trouble signals from the protected premises. When a signal is received, the central station will take such action as is required, such as informing the local fire department of a fire or notifying the police department of the tamper.	Flexible system. Can handle many types of alarms, including trouble, security, intrusion etc. within system at protected premises.
5. Proprietary alarming system	An alarm system that serves multiple buildings under one ownership from a central supervising station at the protected building. Similar to a central station system but owned by the protected property.	Requires attendant round the clock at central supervising station on the premises.
6. Emergency voice/alarm communication system	Includes emergency voice/alarm/or data communications on any of the systems listed above.	Provides personal transmission of information to occupants of the building (including fire department personnel).

rate of information transport that can be carried over an optical fiber than over a wire or a radio channel. (The common word is fiber optics.) All optical conduits and barriers are mechanical objects that conduct or resist the passage of light.

There is no barrier to the passage of an electric or magnetic field, but iron, nickel, and their alloys provide conduits for it which can bypass a space where the field is unwelcome. A system engineer usually wants an energy flow transportation system to offer least resistance so he can get desired pressure, voltage, or flow distribution in building system networks and he wants some means to temporarily close or open energy flow either entirely or partially, either by hand, or by remote control, or automatically. The generic word for such a device in a building energy pathway is "valve," but in common building system lingo, a valve is a device for liquids and gases. (In England, vacuum tubes are called valves.) Following is a list of variety of valves used in building systems including both fluid and other valves so you can see the relationships among them.

1.   Electrical valves include relays, switches, circuit breakers, fuses, vacuum tubes, diodes, SCRs, MOVs, and transistors.

2.   Fluid valves include on–off valves, transfer valves, check valves (fluid diodes), and throttling valves. For on–off, full or no flow requirements, ball, cock, or plug, or gate valves are favored; where tight shutoff is not required, butterfly or slide valves are used; for throttling purpose globe; or for finer adjustment of flow, needle valves are used. They may be operated or adjusted by hand; by electrical, electronic, pneumatic, or hydraulic signals; or by mechanical signals (such as the float-operated valve in a cooling tower). They may be operated by the direction, pressure, or speed of the fluid itself, such as check valves, safety valves, or flow control valves. See Figure 3.17 for different types of valves used in building systems.

3.   Most heat valves are fluid valves for the fluids that carry the heat. Radiant heat valves are made as radiation absorbers or reflectors.

4.   Doors with locks are valves for people. Doors with latches are valves for animals and babies.

3.7   Building Systems Distribution Tree

The distribution tree is the means for delivering information, power, heating, and cooling: the "roots" are the machines, and processors that provide information, power, heat, and cold; the "trunk" is the main duct or pipe that carries air, water, current conductors, optical fibers from the information centers and mechanical/ electrical equipment to the zone to be served; and the "branches" are the many smaller ducts or pipes or fiber nerves that lead to individual spaces.

For now, the questions to be answered about distribution trees for buildings are: How many? What kind? Where? A building can have one giant distribution tree, several medium-sized trees, or an orchard of much smaller trees. At one extreme, a large building system room is the scene of all power, heating, cooling, and information production; leading from this room is a very large trunk of duct and pipes with perhaps hundreds of branches. Power and signal pipes are like tree creepers attached to the trunk and branches of the tree. At the other extreme, each zone is self contained and has its own electrical/mechanical equipment (such as a rooftop heat pump, transformer, panel board, and processor), with short trunks and relatively few branches on each tree.

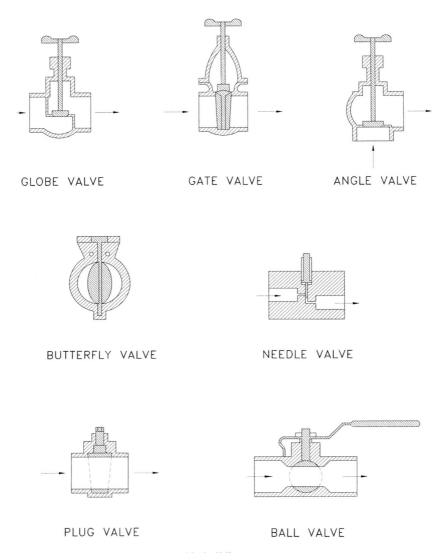

GLOBE VALVE   GATE VALVE   ANGLE VALVE

BUTTERFLY VALVE   NEEDLE VALVE

PLUG VALVE   BALL VALVE

Figure 3.17   Valves commonly used in building systems.

What kind of distribution tree? Most simply, it can be an air (ducts), water (pipes), power (pipes and ladders), or information (fiber conduits and wire pairs) tree. Air distribution trees are bulky and therefore likely to have major visual impacts unless they are concealed above ceilings or within vertical chases. Water and power distribution trees consume much less space (a given volume of water

carries vastly more heat that than does the same volume of air at the same temperature) and can be easily integrated within structural members, such as columns. Further, information distribution trees consume less space and can be supported off any main tree provided they maintain distance to power and water
pipes. Both air and water trees can be sources of noise.

Distribution trees are rich in form possibilities; their variations are enormous.
For now, assume that the choices of how many, what kind, and where have tenta-
tively been made and that the details and variations are to be considered later.
The designer who has chosen the basic form, type of envelope components, and
a support system distribution tree with different branches can now consider questions of size.

Another important factor in system choice is the amount of space the system
requires. In some cases, it is easy to provide small building system equipment
rooms at regular intervals throughout a building, such that little or nothing in the
way of a distribution tree will be required. In other cases, a network of distribution trees and central, large equipment spaces are easier to accommodate.

The relocation of risers at the perimeter (where they are most needed) provides a reduction in the length—and consequently in the size—of pipe and duct
branches. This avoids the need to move large assemblies of building "arteries
and veins" to the surfaces they serve and eliminates bulky ceiling crossovers. It
is logical to place at the perimeter the parts of the system that deal with the effects of sun, shade, and temperature change in the different perimeter zones,
leaving at the core a separate network to handle the more stable interior areas.
The disadvantages of perimeter distribution include (usually) higher construction
costs and an environment that is more thermally hostile, owing to the extremes
of outdoor temperature.

The integration of distribution tree with building structures is required owing
to increases in the complexity and size of the building distribution systems with
technological development (typically, in modern buildings more air and power is
required to power and cool a space owing to high internal loads). The size of
structural systems is reduced owing to the increased strength of materials. The
"uncluttered" floor areas between the more widely spaced columns are now desirable for flexibility in spatial layouts. With the building system branches at or
within these columns, floor areas remain clear, thus giving building system structural integration further impetus. With the high efficiency expectations for cooling, the refrigeration cycle's cooling tower often moved to the roof, taking the
bulky air-handling machinery with it. This further encourages the merging of
systems, for one system is growing wider as the other is diminishing.

Thus, while it is possible to wrap the building systems in a structural envelope, it is of questionable long-term value, given the differing life spans and
characteristics of these systems. The functions of these systems differ widely;
compared to the dynamic on–off air, water, and electrical distribution systems,
the structural system is static. Changes in occupancy can mean enormous

changes in systems, requiring entirely different equipment; structural changes of such magnitude usually occur only at demolition. Building systems can invite user adjustment; structural systems rarely do.

To overcome the above flexibility issues *horizontal distribution* above corridors is very common, since reduced headroom here is more acceptable than in the main activity areas. Further, corridors tend to be away from windows, so their lower ceilings don't interfere with daylight penetration. Since corridors connect nearly all spaces, horizontal service distribution to such spaces is also provided. Furthermore, exposure of these services above corridors can heighten the contrast between such serving spaces and the uncluttered, higher ceiling offices that are served.

Rooftop equipment and distribution tree offers the *economizer cycle* which uses cool outdoor air, as available, to ease the burden on a refrigeration cycle as it cools the recirculated indoor air. The economizer cycle can thus be thought of as a central mechanical substitute for the open window. Relative to open windows, this cycle has several advantages: energy-optimizing automatic thermal control, filtering of fresh air, tempering of the cool outdoor air to avoid unpleasant drafts, and an orderly diffusion of fresh air throughout the building. Its disadvantages are the loss of personal control that windows offer and little awareness of exterior–interior interaction.

3.7.1 Tree Care Cost and Conservation

One big advantage of central equipment rooms is the opportunity they present for energy conservation. Regular maintenance is simplified when all the equipment lives in a generous space kept at optimum conditions; with regular maintenance comes increased efficiency of operation. Another conservation opportunity is that of heat transfer between various machines, or between distribution trees, where one's waste meets another's need.

Within equipment, heat transfer can occur in a *boiler flue economizer,* through which the hot gases in a boiler's stack are passed for use in preheating of the incoming boiler water. For cooling equipment, *dual-condenser chillers* can reject their heat either to a cooling tower (via the heat rejection condenser) or to the building (via the heat recovery condenser).

There are numerous methods for heat transfer between distribution trees—especially between building exhaust air and fresh air. When these two air streams are rather far apart, a set of *runaround coils* can be used. *Energy transfer wheels* go further, in that they transfer latent as well as sensible heat. In winter, they recover both sensible and latent heat from exhaust air; in summer, they can both cool and dehumidify the incoming fresh air. Seals and laminar flow of air through the wheels prevent mixing of exhaust air and incoming air. We commonly experience daily changes from warmer to colder conditions, even in summer. Thermal storage at the base of equipment tree for large buildings can take advantage of this cycle to increase operating efficiency and save energy.

To carry the tree analogy to its logical conclusion, in the spaces served consider the "leaves," the points of interchange between the piped and ducted heating or cooling; cabled or conducted power and lighting; wired or fibered information and signals; and piped or hosed fire suppression and drinking water. In return leaves shade, comfort, nourish, and protect the spaces served.

4

Integration of Mother Nature and Building Systems

4.1 Introduction

A designer's early decisions in site planning will influence the later choices of both the building's mechanical and electrical equipment and its overall consumption of energy. If the site is seen as a collection of resources (sun, wind, water, plants) and also as a part of the environment we all share, the buildings we design can approach self-sufficiency of energy supply, without limiting the availability of local energy resources for neighboring buildings. In the interest of minimizing life cycle cost attributed to the building environmental control system, the following areas should be addressed in planning and design of building systems:

- Site selection and building positioning
- Exploitation of natural geographical assets
- Utility selection
- Fuel selection
- Building architecture, type of construction, and materials
- HVAC equipment selection and operation

Building system loads can be minimized by using nature and site-integrated architectural designs, building materials, and construction techniques that are matched to the local features of the site. Building designs should consider the following items:

- Site orientation to take maximum advantage of terrain features

- Architectural design appropriate for the geographical location
- Control and use of solar radiation
- Protection from adverse winds
- Use of insulation and vapor retarders
- Intelligent selection of HVAC equipment and integration of systems.

It is becoming more common to adapt architectural design to accommodate building utility systems, rather than "fitting" utilities into available space. The use of reinforced construction techniques to increase structural strength against predicted wind pressures can also improve insulating values and building tightness against heat gains and losses. Core areas of concrete floors, walls, and ceilings can be filled with insulating materials to increase thermal resistance ($°F \cdot ft^2 \cdot h/Btu$ or m^2r/w), R-values. Increased use of concrete also provides excellent opportunities for application of thermal storage energy-conservation techniques. For example, large concentrations of concrete or masonry can be used as a solar energy storage mass in active and passive collection systems.

4.2 The Atmosphere Surrounding Buildings

The atmosphere of the earth is defined as "the gaseous mass or envelope surrounding earth, and retained by the earth's gravitational field." The atmosphere surrounding buildings is full of gases, water vapor, water, ice, and heat/warmth of various types. See Table 4.1 for a detailed composition of the atmosphere.

Water is not simply warm ice and water vapor is not just warm water, but in all its phases water is the same substance, H_2O. At a phase change the molecules acquire sufficient energy to move more freely. Ice consists of molecules locked together. If they can acquire enough energy, they will separate and move freely in relation to one another, and the solid will turn into liquid. If the molecules of the liquid acquire still more energy, they will fly off by themselves into the air, and the liquid will turn into a gas. At each change, additional energy—as heat—is needed to give each molecule the necessary freedom to break free, to break the bonds that determine the distance between it and its neighbors. This is called "latent" heat, because all it does is change the phase.

The temperature of the water—at the freezing or boiling point—is not altered at all by the absorption of latent heat. When the phase changes in the reverse direction—and water vapor condenses or water freezes—the same amount of latent heat is released. Different compounds absorb and release different amounts of latent heat. In the case of water, at freezing point, it takes 1076 BTUh (271,000 calories) of heat to vaporize 1 pound (2,500,000 joules/kg), and 144 BTUh (36,255 calories) to melt 1 pound of ice (335,000 J/kg). In a cloud, the latent heat released by the condensation of water vapor is imparted to the sur-

Table 4.1 Composition of the Atmosphere and required quality of Indoor air

Component	Concentration By Volume (Parts per unit volume)	Residence Time (years)	Indoor Air Quality Safe Limit (mg/m^3)
N$_2$ (nitrogen)	0.78084	$4 \cdot 10^8$ y for cycling through sediments	
O$_2$ (oxygen)	0.20946	6000 y for cycling through biosphere	
H$_2$O (water)	$(4-0.004) \cdot 10^{-2}$	—	
Ar (argon)	$9.34 \cdot 10^{-3}$	Largely accumulating	
CO$_2$ (carbon dioxide)	$0.346 \cdot 10^{-3}$	10 y for cycling through biosphere	9000
Ne (neon)	$1.818 \cdot 10^{-5}$	Largely accumulating	
He (helium)	$5.24 \cdot 10^{-6}$	$2 \cdot 10^6$ y for escape	
CH$_4$ (methane)	$1.55 \cdot 10^{-6}$	2.6–8 y	
Kr (krypton)	$1.14 \cdot 10^{-6}$	Largely accumulating	
H (hydrogen)	$5.5 \cdot 10^{-7}$	4–7 y	
N$_2$O (nitrous oxide)	$3.3 \cdot 10^{-7}$	5–50 y	30
CO (carbon monoxide)	$(2-0.6) \cdot 10^{-7}$	0.5 y	29
Xe (xenon)	$8.7 \cdot 10^{-8}$	Largely accumulating	
O$_3$ (ozone)	$(3-1) \cdot 10^{-8}$	—	0.05
CH$_2$O (formaldehyde)	$<1 \cdot 10^{-8}$	—	0.37
NH$_3$ (ammonia)	$(20-6) \cdot 10^{-9}$	About 1 d	
SO$_2$ (sulfur oxide)	$(4-1) \cdot 10^{-9}$	Hours to weeks	5.2
NO + NO$_2$ (nitrogen oxides)	10^{-9}	<1 mo	5.6
CH$_3$Cl (methyl chloride)	$5 \cdot 10^{-10}$	—	
CCl$_4$ (carbon tetrachloride)	$(2.5-1) \cdot 10^{-10}$	—	31
CF$_2$Cl$_2$ (Freon 12)	$2.3 \cdot 10^{-10}$	45–68 y	
H$_2$S (hydrogen sulfide)	$\leq 2 \cdot 10^{-10}$	$\leq$1 d	
CFCl$_3$ (Freon 11)	$1.3 \cdot 10^{-10}$	45–68 y	

Concentrations, by volume, of components and their residence times are given at ground level (excluding local pollutants). Indoor air quality safe limits are based on American Industrial Hygiene Association, based on 8 hr/day, 5 days a week exposure.

rounding air which is warmed. This also explains why it often feels warmer when snow falls after a period of intensely cold weather, and why the spring thaw produces a distinct chill.

Psychrometrics deals with thermodynamic properties of moist air and uses these properties to analyze conditions and processes involving moist air. The amount of water vapor air can hold depends on the air temperature. Not only can warm air hold more water vapor than cool air, but its capacity to do so increases with rising temperature at an accelerating rate. At 50 °F (10 °C), for example, a cubic foot of air can hold 0.009 ounces of water vapor (9.4 g/m³). At 68 °F (20 °C), a cubic foot can hold 0.02 ounces (17.3 g/m³), and at 86 °F (30 °C) it can hold 0.03 ounces (30.4 g/m³). As the temperature increases from freezing to 50 °F (10 °C), the water vapor capacity of a cubic meter of air increases by 0.16 ounces (4.6 g), but an increase from 77 °F (25 °C) to 95 °F (35 °C)—in both cases a rise of 18 °F (10 °C)—increases the water vapor capacity by 0.58 ounces (16.55 g), about three and a half times more.

The temperature at which a volume of air is fully saturated with water vapor is called its "dew-point" temperature. If air is cooled still further, in most cases (but air can be supersaturated) the vapor will start to condense. Obviously, the dew-point temperature depends on the water vapor content of the air—the drier the air, the lower the dew-point temperature will be. The amount of water vapor in the air, expressed as a percentage of the amount needed to saturate the air at that temperature, is the "relative humidity" of the air, and it reaches 100 percent at the dew-point temperature. The relative humidity is the figure most commonly quoted in weather reports that give the humidity, but meteorologists also use other measures. The "absolute" humidity is the actual mass of water vapor present in a given volume of air (usually 1 m³), and the "specific" humidity is the mass of water vapor present in a given mass of air (usually 1 kg). It is possible for the dew-point temperature to be below freezing. In this case the vapor changes directly into ice without passing through the liquid phase, and the clouds are composed of ice crystals rather than water droplets.

The temperature of the air decreases as altitude increases. The rate of change is called the "lapse rate," and its average value is about 3.6 °F per 1000 feet (6.5 km). Air is a poor conductor of heat but it is very mobile. This means that although its molecules are very inefficient at passing energy from one to another by direct contact, they are much more efficient at transferring heat by convection. When air is warmed by contact with the surface, it rises. If you picture a "parcel" of air, as it rises it will cool adiabatically—fairly rapidly while it remains unsaturated and more slowly as its water vapor starts to condense and releases latent heat to offset the cooling. Adiabatic cooling and warming involve no mixing between the "parcel" of air and the air surrounding it. This is the way rising air cools once it is well clear of the surface—although there is usually some mixing with the surrounding air—but close to the surface it is more usual for air to be mixed fairly thoroughly by local turbulence as it moves laterally across the uneven ground surface.

The earth is warmed by the sun more strongly at the equator than at the poles. This suggests the possibility of a quite simple convection system. In low latitudes in each hemisphere, air is made very warm. It rises and its place near the surface is taken by cooler, denser air drawn in from a higher latitude. The rising air cools adiabatically and, somewhere near the pole, it is drawn downwards again to replace air that has moved equatorwards. Throughout the whole of the troposphere, the air is engaged in a kind of rolling motion. If this is all that happened, we should expect water vapor, carried upwards in the warm, low-latitude air, to condense and fall as rain not very far from the equator. In higher latitudes the air would be very dry—"squeezed" dry as the air approached the tropopause—and so the climate of the tropics would be very humid and that of high latitudes very dry. As we know from our own experience of the weather from one day to the next, it is not so simple!

The oversimplification begins with our implied assumption that the warming of the surface occurs evenly. It does not. Nearly three-quarters of the surface of the earth is covered by water. Like air, water is a poor conductor of heat. It also has high "specific heat." This is the amount of heat energy that a given volume of a substance must absorb to increase its temperature by a specified amount. Because its specific heat is high, it takes a considerable amount of solar radiation to alter its temperature. Unlike air, water is warmed from above, so a surface layer of warm water overlies the main water mass, which is much cooler. Oceanfront power plants produce energy by exploiting this temperature difference (about 40 °F) between warm surface and cold ocean depths below 2000 feet. The two mix, transporting heat downwards from the surface, and the consequence is that the oceans warm and cool slowly. Rocks, on the other hand, have a specific heat about one-fifth that of water. They, and soil particles, which are made from rock fragments, warm and cool quickly and heat can be transferred by conduction among particles that are in contact with one another. There is a very marked difference, therefore, between the effect of surface warming over land and over water.

There is also a difference between one soil and another. If the soil consists of large rock particles, loosely packed with air spaces between them, the air will act as an insulator. The surface will warm rapidly during the day, but its heat will not be conducted to any great depth because of the air. Sand is composed of large mineral particles of this kind. Where the particles are smaller, as in a clay soil, for example, particles will be more closely packed, with smaller air spaces, and heat will be conducted more effectively, so the soil warms quickly, but also absorbs more heat than a sandy soil. In a moist soil, spaces between solid particles will tend to be filled, and the soil will be a better conductor. If the soil is very wet, however, the high specific heat of the water will become the dominant factor, and the soil will heat and cool more like water than like dry land. Such variations, some very local and others vast, influence the way the building systems and the envelope system surface respond to the warmth they receive.

4.3 Winds Surrounding Buildings

Wind is air that is moving in relation to the earth's surface, and it moves because of differences in air pressure in the atmosphere. Without these differences, no wind would blow. Differences in pressure develop over areas where the sun heats the earth's surface unevenly. Wherever the earth is warmer, air heats up and expands, and air pressure increases, compared to the pressure over cooler places.

Air can be imagined as lying over the earth in layers between constant-pressure surfaces, with the densest layer at the bottom. Sometimes the air is still and the layers even and flat. But when one area absorbs more heat, the air expands, air pressure rises, and the air pressure layers expand, too, and become curved. Air then begins moving from the high-pressure area to the low-pressure area, producing a wind high above the ground. The greater the temperature difference—and therefore the pressure difference—between two places, the stronger the wind that blows between them.

In high- and low-pressure systems, a centrifugal force also determines wind direction. In the upper atmosphere, the pressure gradient force, turning force, and centrifugal force balance when winds blow clockwise around high-pressure and counterclockwise around low-pressure systems. At ground level, friction turns the winds outward in a high and inward in a low-pressure system.

Breezes: Winds are fairly unpredictable, but some breezes blow as regularly as the rising and setting of the sun. Unlike most winds, these breezes depend more on local changes in temperature than on the presence of large-scale high- and low-pressure systems. Near an ocean or large lake, for example, sea and land breezes blow because the sun warms land faster than it does water. At sunrise, the sun shines on both water and land, but only the temperature of the land rises noticeably. The warmed land heats the air above it. As that warm air rises, it draws colder ocean-level air inland, creating a sea breeze. Shortly after daybreak, the air temperature over sea and shore is equal. With no difference in temperature, convection cannot take place. Once the sun sets, the shore begins cooling and so does the air above it. Before long, the air temperatures over the ocean and shore are the same. The sea breeze stops blowing. This is the *evening calm.* Thus, there is no wind. By the same token, land loses heat faster than water when the sun stops shining. As night progresses, the shore becomes cooler than the surface of the water. The air over the ocean is then warmer compared to that over land, so it rises. Cooler air from the shore then moves out toward the sea, creating a sea breeze.

In a valley, the air rises and sinks over the course of one day as the surrounding mountain sides gain and lose heat. The updrafts and downdrafts that are created by this temperature change produce valley and mountain breezes that blow virtually every day the sun is out.

4.4   The Building Envelope

Just as a satellite's orbit represents a delicate balance between inertia and gravity, a building envelope maintains a delicate balance between surroundings and internal systems. In the case of a satellite, if it were not for gravity, the satellite's inertia would send it straight out of the earth's orbit and into space. Similarly in buildings, if it were not for the integrity of the envelope, the atmospheric wind and radiation inertia would throw the system into disarray and energy out of building.

The envelope of a building is not merely a set of two-dimensional exterior surfaces; it is a transition space—a theater where the interaction between outdoor forces and indoor conditions can be watched. Some of these interactions include the ways in which sun and daylight are admitted or redirected to the interior, the channeling of breezes and sounds, and the deflection of rain. The amount of sun or breeze admitted through a window, or the amount of building heat lost through it, has less often been cited as an influence on window design. The more suited the outdoors to comfort, the more easily indoor activity can move into this outdoor transition space.

The envelope also has a fourth dimension; it changes with time. Seasonal and load changes are a valued component for designers who appreciate that climate and functions change, and that loads are unpredictable. As much as a designer might prefer that a building have a certain identity, unchanging in its expression of a set of ideas, time will pass and changes will occur. Seasonal changes/load switches allow the designer to give control of the building to its users. If the designer has carefully integrated the range of choices that the switch should provide, successful user control will be possible.

4.4.1 Passive Environmental Systems

Radiative cooling is a natural heat loss mechanism that causes the formation of dew, frost, and ground fog. Because its effects are the most obvious at night, it is sometimes termed nocturnal radiation, although the process continues throughout the day. Thermal infrared radiation, which affects the surface temperature of a building wall or roof, has been treated in an approximate manner using the solar–air temperature concept. Radiative cooling of window and skylight surfaces can be significant, especially under winter conditions when the dew-point temperature is low.

The most useful parameter for characterizing the radiative heat transfer between horizontal nonspectral emitting surfaces and the sky is the sky temperature T_{sky}. The sky temperature is a function of atmospheric water vapor, the amount of cloud cover, and air temperature. The lowest sky temperatures occur under an arid, cloudless sky. The monthly average sky temperature depression, which is the average of the difference between the ambient air temperature and the sky temperature, typically lies between 9 and 43 °F. In regions where sky temperatures fall below 61 °F 40% of the month or more, all nighttime hours are effectively available for radiative cooling systems.

Passive Heating Systems: Every building is passive in the sense that the sun tends to warm it by day and it loses heat at night. Passive systems incorporate solar collection, storage, and distribution into the architectural design of the building and make minimal or no use of fans to deliver the collected energy to the structure. Passive solar heating, cooling, and lighting design must consider the building envelope and its orientation, the thermal storage mass, and window configuration and design.

Direct gain passive system uses a large expanse of south-facing glass to admit solar radiation, whereas indirect gain solar houses use the south-facing wall surface or the roof of the structure to absorb solar radiation, which causes a rise in temperature that, in turn, conveys heat into the building in several ways.

By glazing a large south-facing, massive masonry wall, solar energy can be absorbed during the day, and conduction of heat to the inner surface provides radiant heating at night. The mass of the wall and its relatively low thermal diffusivity delays the arrival of the heat at the indoor surface until it is needed. The glazing reduces the loss of heat from the wall back to the atmosphere and increases the collection efficiency of the system.

Openings in the wall, near the floor, and near the ceiling allow convection to transfer heat to the room. The air in the space between the glass and the wall warms as soon as the sun heats the outer surface of the wall. The heated air rises and enters the building through the upper openings. Cool air flows through the lower openings, and convective heat gain can be established as long as the sun is shining. Buildings whose skins are plentifully supplied with switches become continuing demonstrations that architecture is a performing art, not just sculpture.

4.5   Integration of Atmosphere and Building Systems

The use of on-site resources not only can reduce the amount of energy needed to maintain the interior climate, but it also can produce outdoor spaces that become especially pleasant to use. Such spaces can direct winter sun to a glass wall while blocking the wind, or funnel the summer breeze through shade to an open window. Site planning is greatly influenced by economic considerations, zoning regulations, and adjacent developments, all of which can interfere with the design of a site to utilize the atmospherical sun, rain, plants, and the wind. *Integration* of all these concerns at the site-planning stage is the first step in adapting a building to its climate.

Schematic site plans are typically used as a kind of inventory; overlaid "bubble diagrams" can test possible design integration that can relate rooms and functions to their surroundings in the plan. Sun and wind conditions (in both summer and winter), noise sources, and water runoff patterns can be included in this schematic plan. It is particularly important to identify microclimates on the site, the places that have special characteristics differing from the regional climate. Microclimates can present opportunities where less energy is consumed because the winter is warmer, or the summer cooler. Or microclimate can be problem areas to be avoided for building or outdoor activity, if possible, or where special design measures need to be taken to correct their difficulties.

Access to Light and Sun: The amount of solar energy available to each site varies both seasonally and daily. The value of daylight (and air) to buildings has long been recognized in zoning laws, which require that minimum distances (setbacks) be maintained between a building and the property line in lower density areas. Height restrictions often accompany these setbacks, defining a maximum "build-able volume" in which a building can grow.

When *direct sun* in winter is desirable at the ground floor of each site, this pyramid changes shape and its volume decreases; this is due to the low angle of the winter sun, which is readily blocked by taller objects south of a given window. This most restricted pyramid (called the "solar envelope") is at present rarely achieved, but various proposals to guarantee access to direct winter sun for solar collection are under development at federal, state, and local levels.

Sound and air are considered together because they are so difficult to separate. Many buildings that could be opened to ventilation or cooling by breezes rely instead on forced ventilation because of the noise that would accompany the breeze through an open window. Polluted air is another potential deterrent to "natural" ventilation. Almost any device that reduces sound will also reduce the velocity of the breeze, as is true of most filtering devices to remove dirt particles.

Wind Control: For most buildings, wind (like sun) changes from friend to enemy with the change of seasons. Control of wind often means utilizing wind-sheltered areas in winter, while encouraging increased wind speeds for the building in the summer. The pressure differences, flow patterns, and the size and shape of wind-protected areas behind the obstacle are all utilizable for control of air motion inside and out, in summer. While one of the deterrents to operable windows is the dirt that they could admit to the interior, such air pollution is a threat to people and their buildings in other ways as well. Buildings (and the electric generating plants that supply electrical energy to them) are major contributors to the pollution of the "fresh air" needed for ventilation.

The building form can work with the climate to produce air motion for cooling, although the faster air speeds that extend the human comfort zone above 83 °F (28 °C) may be difficult to provide without mechanical assistance. Relative humidity is still controlled by mechanical (or chemical) means. Building form and materials may be able to keep spaces surprisingly cool, but without mechanical or chemical dehumidification surfaces become clammy and covered with mold. Also it is difficult to filter air without a fan to force the air through the filtering media. Thus, whereas the desired air and surface temperatures can often be achieved by passive means (a combination of building form, surface material, and occasional user response), the comfort determinants of air motion, relative humidity, and air quality often require mechanical devices.

As the control of air properties—motion, moisture, particulate content—becomes more critical to comfort, the designer becomes more likely to respond with a sealed building, excluding outdoor air except through carefully controlled mechanical equipment intakes. In the recent past, this exclusion of outdoor air has often resulted in exclusion of daylight, of view, of solar heat on cold days— in sum, by a general rejection of all aspects of the exterior environment. As designers come to terms with the role of mechanical and electrical equipment, we should also clarify the role of these devices in relation to the climate: Are they occasional modifiers, permanent interpreters, or permanent excluders of the outdoors?

When daylight and wind ventilation are desired, they combine to limit the width of buildings. This is particularly evident in multistory office buildings, where increasing urban density and reliance on electric lighting and cooling have changed the form of buildings considerably in the past few decades. The greater the depth of a given room, the more that room's interior requires cooling throughout the year. The ratio is particularly notable in "intelligent" office buildings. On the other hand, such offices require heating around the perimeters in winter if the insulation of the windows is poor. This causes a cross-mixing of warm and cool air in the space between the perimeters and the interiors, resulting in a substantial waste of energy. To avoid such waste, air flow windows can be provided and achieve a perimeterless air conditioning system that eliminates cross-mixing. Figure 4.1 shows the air flow windows concept using room air to wash off the exterior window surface to avoid condensation.

The most desirable balance among the energy needs for heating, cooling, and day lighting will vary both by building type and by climate. The thermal zoning of a building recognizes that different envelope and support systems may be required within the building. Figure 4.2 shows psycometrics of air inside building for a typical cooling and heating process. The more carefully zoning is considered in these early design stages, the better will be the thermal performance and the lower will be the annual energy consumption. (Also, the less likely it will be that all sides of a building will have an identical appearance.)

Rain and Groundwater: Most buildings interact with water in four forms: rainwater, groundwater, potable, and fire protection water (brought to and taken from urban sites by utilities). Like solar energy, *rainwater* is a diffuse, intermittent, and often seasonal resource. As a source of water, it is most often collected and used where other water sources are scare, or of poor quality. Rain also has an influence on building design: heavy rains and pitched roofs have long been found in the same locales. Overhangs may extend further beyond walls exposed to storm winds; even gutter and downspout details can become a design feature.

The ability of water to conduct and store heat, which encourages its use as a heat sink, also makes it a thermal enemy of heat storage tanks or bins in solar buildings. There is little point in collecting solar energy to store in an underground tank for later use if groundwater is allowed to rob the tank of its heat. On the other hand, groundwater helps storage tanks that are to be kept as cool as possible. Such tanks provide cool water to buildings at peak-heat hours of hot days and can greatly lessen the demand for electricity to run conventional air conditioners.

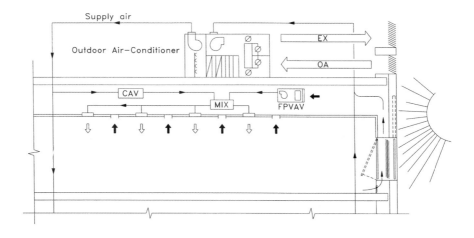

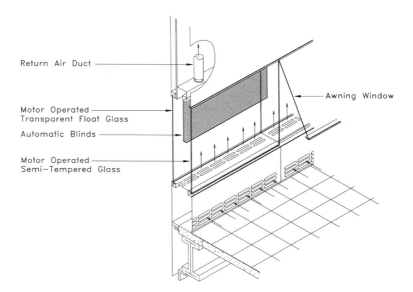

Figure 4.1 (Top) Daylight and wind-integrated intelligent HVAC system of typical office floor. (Bottom) Sectional view of daylighting and air flow window.

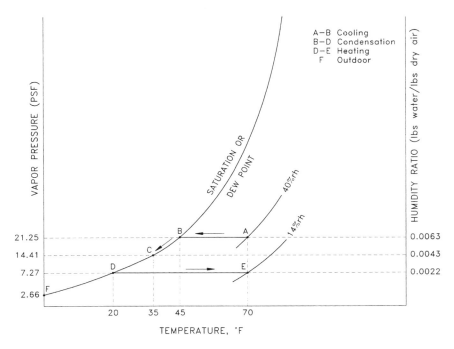

Figure 4.2 Typical cooling and heating processes in building PSF = pounds per square foot.

Plants have several roles: they are part of the water cycle; they turn carbon dioxide into oxygen by day; they provide organic mater suitable for composting (eating, in some cases); and they help us tell time both by growth and by change with seasons. Our associations with plants are mostly pleasant ones, and they contribute to our enjoyment of the places where they grow. Plants are also of immediate practical value to buildings because they enhance privacy, slow the winter wind, reduce the glare of strong daylight, or prevent summer sun from entering and overheating buildings. In this latter role, plants are particularly noteworthy, because they enhance a feeling of coolness when breezes rustle or sway their leaves, and especially because they respond more to *cycles of outdoor temperature* than to those of *sun position*. Unlike sunscreens on buildings, plants thus provide the deepest shade in the hottest weather.

4.6   Energy Resources and Buildings

Energy supplies are classified as either renewable or nonrenewable.

Renewable fuel sources are those that are available indefinitely but arrive at a relatively fixed rate; the influx of solar energy varies from day to day, but on the average it continues at a steady rate. A woodlot will produce a limited amount of wood per year, but it will do so for centuries if properly managed. A popular analogy for renewable fuel sources is a fixed-but-steady monetary income, such as a salary with no raises. Using the same analogy, nonrenewable fuel sources are like savings accounts that draw no interest—once spent, they are gone.

The *nonrenewable resources* we utilize can be bought and used in large quantities all at once, which make possible many processes that are difficult to attain with low-concentration, steady, renewable resources. Thus, the use of nonrenewable fossil and nuclear fuels in high-temperature and high-concentration processes—for example, in power-generating plants—is widespread and is likely to continue until resource depletion is more closely approached.

Electricity is the most prevalent form of energy in a modern building. It supplies not only electric outlets and electric lighting, but also the motive power for ventilation, heating, and cooling equipment; traction power for elevators and material transport; and power for all signal and communications equipment. An electric power failure can paralyze a facility. If properly designed, the facility will return to partial functioning by virtue of emergency equipment that will furnish some part of the facility's emergency needs for a limited time. Given this complete dependence on electric power for normal functioning, it is apparent that system designers in other disciplines must be familiar with normal electrical systems. Most projections of future energy consumption assume continued, though less rapid, growth in demand.

1. Consumption of electricity is expected to rise about twice as fast as the overall energy demand, and more and more we are expected to use electricity in place of other energy forms.

2. Electricity is almost the only source of illumination, driving motors and computers in buildings, other than daylight; and passive heating/cooling.

3. Electricity is a convenient and versatile energy form; it not only serves such high-temperature and highly concentrated (or "high-grade") energy tasks as lighting and motive power, but it can also serve the low-temperature simpler (or "low-grade") tasks, such as cooling, heating water, and space heating for buildings. "All-electric" buildings have become commonplace, even though they are subject to paralysis in blackouts, as is any building dependent on a single energy source.

4. Electricity delivers only about one third the total energy that goes into its production; the other two thirds is usually lost as waste heat at the generating plant.

Thus, the relationships among greater energy demands, electricity, lighting, M/E equipment, and interior climate are complex. The higher-grade tasks for which electricity is suitable also need careful consideration. The annual energy growth rate for the power of building is projected at about 10%. Buildings designed today may outlast the supplies of the fuels that currently support them. With the apparent depletion of nonrenewable energy resources, a "scarcity ethic" has developed. This emphasizes conservation (demand side management) rather than increased production, and brings "user control" of on-site energy sources back among architectural design considerations. The earlier design response, however, was pulled in seemingly opposite directions: a tendency to *close in* the building to conserve energy, versus the notion of *opening out.*

In the next century, reducing the consumption of nonrenewable fuels and increasing the use of renewable resources should continue to be the primary goals of developing building system technology. Buildings will be designed and constructed to make them adaptive to the natural conditions and features of their geographical location. Until recently, building material choices were rarely influenced by the amount of energy invested in those materials. Now that such information is becoming more widely available, energy content can be included with other factors—aesthetics, durability, recycle potential, fire resistance, labor intensiveness, and installed cost—that a designer considers in choosing materials for a building.

The importance of heat mass and fenestration is being given more attention in building construction. Selective use of fenestration will increase daylighting and reduce the heat generated by artificial lighting, as well as lower costs. Fiber optic devices will distribute natural light to interior building spaces, reducing the demand for energy-consuming and heat-producing artificial lighting. Use of solar energy for power, heating and cooling will be expanded. Building systems designers can have a positive impact on society through energy conservation, and buildings that encourage their users to directly experience the natural environment will both facilitate energy conservation and enrich the user's comfort experience.

Increased use of thermal storage will reduce the use of energy resources for both heating and cooling. Available heat energy collected from the sun and recovered from industrial processes will be stored in water and masonry heat sinks and substances with special properties. Meeting the increasing demand for cooling will be aided with the use of supplemental systems, such as ice bank storage.

HVAC equipment will continue to be improved, as efforts to reduce fuel consumption are pursued. New equipment is smaller, cleaner, and more efficient. Stronger and lighter materials permit greater efficiency and innovation in architectural design. Improved insulating materials reduce unwanted heat gains and losses. Fuel burning furnaces and boilers are being made more efficient by such innovations as pulsed burning, exhaust heat recovery, and improved materials. Substances that store and release large amounts of heat during chemical reactions or changes in physical state will find increasing application in the HVAC systems of buildings.

Heat recovery applications will grow as improved heat exchangers are developed. Continued public concern over pollution will drive the development of improved ventilation and filtration equipment. Superconducting substances now being researched will revolutionize power distribution and control circuitry. Finally, the use of computerized devices for the design and operation of environmental control systems will become commonplace.

5

Integration of Solar Power and Building Systems

5.1 Introduction

Sunlight, the energy source that makes all life possible, may someday furnish much of the electricity, heating, and cooling on which modern society runs. To minimize energy consumption in constructing and using buildings, and to integrate them with their surroundings, the conditions best suited to various functions should approach the characteristics of the "solar layer" of the site in which they are located. This possibility provides two major challenges for designers:

1. To design our buildings not only to save energy, but so that they can also eventually be weaned away from dependence on nonrenewable fuels.

2. To use energy wisely; to expect only a "fair share" of locally available renewable fuels, recognizing that such sources are limited even though they are continually available.

For example, with supplies stretched tight by increasing density, it is tempting to erect a larger solar collector to intercept sunlight that would otherwise be utilized by the neighboring building. The temptation grows stronger as the building is designed to rely more heavily on the sun. As we face a continued increase in world population and continued depletion of renewable resources, it seems obvious that labor-intensive alternatives must be made available. The designer's challenge is to make these alternatives attractive and rewarding to use.

Sunlight can be used directly or indirectly to power a building. Indirect power generation uses a set of mirrors to focus the sun's energy onto a heat exchanger, which vaporizes water or some other liquid, producing steam. This steam is then used to drive a conventional turbine and generator. Direct power solar system utilizes silicon solar cells to generate electricity directly from sunlight. Building integrated photovoltaics (BIPV) commonly called solar cells or modules can be installed in building architectural elements such as wall glazing, awnings and skylights. Building system integration is coupling between BIPV and building electrical and mechanical systems. Photovoltaics can power or charge batteries during sun hours that serve as an off hour power source.

5.2   Obstacles

The major obstacles encountered in solar heating and cooling are economic, resulting from the high cost of the equipment needed to collect and store solar energy. In some cases, the cost of the solar equipment is greater than the resulting savings in fuel costs. Some problems that must be overcome are inherent in the nature of solar radiation, for example:

- It is relatively low in intensity, rarely exceeding 300 Btuh/ft^2. Consequently, when large amounts of energy are needed, large collectors must be used.

- It is intermittent because of the inevitable variation in solar radiation intensity from zero at sunrise to a maximum at noon and back to zero at sunset. Some means of energy storage must be provided at night and during periods of low solar irradiation.

- It is subject to unpredictable interruptions because of clouds, rain, snow, hail, or dust.

- Systems should be chosen that make maximum use of the solar energy input by effectively using the energy at the lowest temperatures possible.

The performance of any solar energy system is directly related to the (1) heating load requirements, (2) amount of solar radiation available, and (3) solar energy system characteristics. Various calculation methods use different procedures and data when considering the available solar radiation. Some simplified methods consider only average annual incident solar radiation, while complex methods may use hourly data.

The cost-effectiveness of a solar domestic and service hot water heating system depends on the initial cost and energy cost savings. A major task is to determine how much energy is saved. The annual solar fraction—the annual solar contribution to the water heating load divided by the total water heating load—can be used to estimate these savings. It is expressed as a decimal fraction or percentage and generally ranges from 0.3 to 0.8 (30% to 80%), although more extreme values are possible.

5.3 Active Solar System

Active systems must have a continuous availability of nonrenewable energy, generally in the form of electricity, to operate pumps and fans. A complete system includes solar collectors, energy storage devices, and pumps or fans for transferring energy to storage or to the load. The load can be space cooling, heating, or hot water. Although it is technically possible to construct a solar heating and cooling system to supply 100% of the design load, such a system would be uneconomical and oversized for seasonal operation. The size of the solar system, and thus its ability to meet the load, is determined by life cycle cost analysis which weights the cost of energy saved against the amortized solar system cost.

5.3.1 *Components*

Active solar system components are the same as those in conventional heating, air conditioning, and hot water systems (pumps, piping, valves, and controls), and their installation is not much different from conventional system installation. Solar collectors are the most unfamiliar component used in a solar energy system. They are located outdoors, which necessitates penetration of the building envelope. They also require a structural element to support them at the proper tilt and orientation toward the sun. The mounting structure should be built to withstand winds of at least 100 mph, which impose a wind load of 40 lb/ft^2 on a tilted roof. Wind load requirements may be higher, depending on local building codes.

The collectors should be located so that shading is minimized and should be attractive both on and off site. They should also be located to minimize vandalism and should be placed as near to the storage tank as possible to reduce piping costs and heat losses. For best annual performance, collectors should be installed at a tilt angle above the horizontal that is appropriate for the local latitude. They should be oriented toward true south, not magnetic south.

5.3.2 *Salient Design Features*

The following list is for designers of solar heating and cooling systems. Specific values have not been included because these vary for each application.

Collectors: Ensure that collector area matches application. Review the collector instantaneous efficiency curve and match between collector and system requirements. Relate collector construction to end use; two cover plates are not required for low-temperature collection in warm climates and may, in fact, be detrimental. Two cover plates are more efficient when the temperature difference between the absorber plate and outdoor air is high, such as in severe winter climates or when collecting at high temperatures for cooling. Flat black surfaces are acceptable and sometimes more desirable for low collection temperatures.

Will snow hang up and ice formation take place? Will casing vents become blocked? Do collectors on roof present rainwater drainage or condensation problems? Do roof penetrations present potential leak problems? Will materials deteriorate under operating conditions? Will any pieces fall off? Are liquid collector passages organized in such a way as to allow natural fill and drain? Does the air collector duct connection configuration promote a balanced airflow and an even heat transfer? These are a few system installation pitfalls that should be checked.

Freeze protection is important and is often the determining factor when selecting a system. Freezing can occur at ambient temperatures as high as 42 °F because of radiation to the night sky. Freeze protection is provided by using fluids that resist freezing. Fluids such as water/glycol solutions, silicone oils, and hydrocarbon oils are circulated by pumps through the collector array and double-wall heat exchanger.

Hydraulics: If antifreeze is used, check that the flow rate has been modified to allow for the viscosity and specific heat. If system uses drain-back freeze protection, check that

1. Provision is made for drain-back volume and back venting.
2. Pipes are graded for drain back.
3. Solar primary pump is sized for lift head.
4. Pump is self-priming if tank is below pump.

Ensure that the collector pressure drop for the drain-back system is slightly higher than static pressure between the supply and return headers. Optimum pipe arrangement is reverse return with collectors in parallel just like conventional chilled or hot water systems. If heat exchangers are used, check that the approach temperature differential has been recognized in the calculations and adequate provisions are made for water expansion and contraction. As a general rule, simple circuits and controls are better.

Airflow: Check that air velocities are within the system parameters and that cold air or water cannot flow from collectors by gravity under "no-sun" conditions. Check duct configuration for balanced flow through collector array; more than two collectors in series can reduce collection efficiency.

Storage: Energy collected from solar heat during sun-up hours needs to be stored for use during sun-down hours. Thermal storage capacity should match the parameters of the collector area, collection temperature, utilization temperature, and system load and thermal inertia should not impede effective operation. Provide for temperature stratification during both collection and use. If liquid storage is used for high temperatures (above 200 °F), check that tank material and construction can withstand the temperature and pressure, storage location does not promote unwanted heat loss or gain, and that adequate insulation is provided. Also protect liquid tanks from exposure to either an over pressure or vacuum.

Controls: Collector loop controls include solar input, collector temperature, and storage temperature. Ensure that controls allow both the collector loop and utilization loop to operate independently. Controls should be fail-safe and control sequences should always revert to the most economical mode of operation.

5.4   Building System Applications

Domestic Hot Water Systems: Domestic hot water loads include short periods of high draw interspersed with long dominant periods. Make provisions to prevent reverse heating of the solar thermal storage by the domestic hot water backup heater (see Figure 5.1). On days of low solar input the system should provide cold makeup water preheating. On days of high solar input the tempering valve should limit domestic hot water supply to a safe temperature. If collectors are used to heat water directly, provide treatment for preventing scale formation in absorber plate waterways.

Warm air heating systems have the potential of using solar energy directly at moderate temperatures. See Figures 5.1 and 5.2 for arrangement of the components. At times of low solar input, solar heat can still be used to meet part of the load by preheating return air. Baseboard heaters require relatively high supply

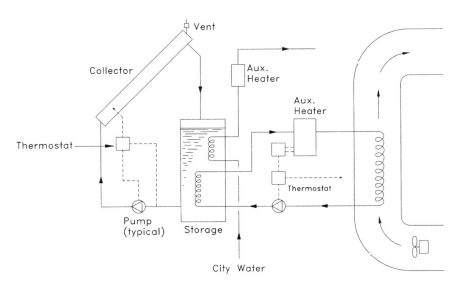

Figure 5.1   Solar collection, storage, and distribution system for domestic hot water and space heating. (Reprinted by permission of the American Society of Heating, Refrigerating and Air Conditioning Engineers, Atlanta, Georgia, from the *1995 ASHRAE Handbook—Applications*)

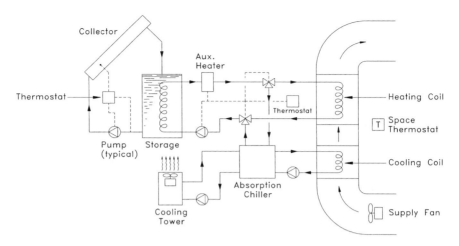

Figure 5.2 Space heating and cooling system using lithium bromide–water absorption chiller. (Reprinted by permission of the American Society of Heating, Refrigerating and Air Conditioning Engineers, Atlanta, Georgia, from the *1995 ASHRAE Handbook—Applications*)

temperatures for satisfactory operation. Their output varies as the 1.5 power of the log mean temperature difference and falls off drastically at low temperatures. If solar is combined with baseboard heating, check that supply temperature is compatible with heating load. Water-to-air heat pumps rely on a constant solar water heat source for operation, and provide backup storage for sun-down periods.

Cooling systems: Solar activated absorption cooling with gas fuel backup is currently the only commercially available active cooling (see Figure 5.2). Be assured of all design criteria and a large amount of solar participation. Available storage capacity can be used for storing both hot water and chilled water.

6

Integration of Daylighting and Building Lighting

6.1 Introduction

Lighting in commercial buildings consumes 25% to 60% of the electric energy utilized. Any attempt to reduce this must necessarily include integration of the cheapest (insofar as energy is concerned), most abundant, and in many ways, most desirable form of lighting available—daylight. This is most applicable at the perimeters of buildings; yet, even the interiors of low-rise buildings can be served with daylight for the general or overall illumination, using small individually controlled high-efficiency electric lights only where and when needed. As with many such substitutions, the designer must consider the tradeoff, will more glass required to admit daylight produce greater heat loss on winter nights and undesired heat gain on summer days? Techniques such as protecting glass against heat loss with insulating shutters and designing windows to minimize summer solar gain need to be evaluated.

Since daylight varies with earth rotation, atmospheric cover, season, and time of the day, it creates special problems of glare control, direct sunlight control, and heat gain limitation. In large measure the science (and art) of daylighting is not so much how to provide daylight as how to do so without the attendant undesirable effects. Light itself is predictable, that is, it obeys certain laws and exhibits certain fixed characteristics.

Lighting "quality" is a term used to describe the overall lighting scene, that is, the luminance, diffusion, uniformity, and chromaticity of the lighting. For example, if two identical rooms are lighted to the same *average* illuminance, one with a single bare bulb and the second with a luminous ceiling, there is a vast difference in the lighting quality of the rooms. Further, there is an observed relationship between the color of the *light* in a space and the range of acceptable levels of illumination. Well-known psychological effects of colors are the coolness of blues and greens and the warmth of reds and yellows. Warm and saturated colors produce activity, conversely cool, unsaturated colors are conducive to meditation. Cool colors also seem to shorten time passage and are well applied in areas of dull repetitive work.

6.2 Lighting Integration Issues

Effective daylighting design requires consideration of the interaction of building systems. For instance, the installation of energy-conserving dimmable lighting systems may provide only minimal savings if heavily tinted glazing (used to control glare) excessively reduces the admission of daylight.

For many years an artificial division existed in the field of lighting design, dividing it into two disciplines: architectural lighting and utilitarian design. The former trend found expression in design that took little cognizance of visual task needs and displayed an inordinate penchant for wall washers and architectural lighting elements. The latter trend saw all spaces in terms of cavity ratios and performed its design function with foot candles, lumens/watt, and dollars as the ruling considerations. That both these trends have in large measure been eliminated is due largely to the efforts of thoughtful architects, building system engineers, and lighting designers, assisted in part by the energy consciousness and advent of electronic control and efficiency in lighting design. The latter spurred research into satisfying real vision needs within a framework of minimal energy use.

The responsible building system designer must consider quantitatively:

1. Daylight—its introduction and integration with artificial light
2. The interrelationship between the energy aspects of artificial and natural lighting, heating, and cooling
3. The effect of lighting on interior space arrangement and vice versa
4. The characteristics, means of generation, and utilization techniques of artificial lighting
5. Visual needs of specific tasks
6. The effects of brightness patterns on visual activity.

Qualitatively, the designer must consider:

1. The location, interrelationship, and psychological effects of light and shadow, that is, brightness patterns
2. The use of color, both of light and of surfaces, and the effect of eliminant sources on object color

3. The artistic effects possible with color and patterns of light and shadow including the changes inherent in daylighting, and so on.

6.3 Daylighting and Shading Materials

The materials and shading systems used in daylighting are characterized by the following properties: transmission, absorption, reflection, diffusion, refraction, cost, thermal properties, and appearance.

High-Transmittance Materials: These transmit light without appreciably changing its direction or color; they are image preserving. Common types are sheet, polished plate, and float and molded glass as well as some rigid plastic materials and formed panels. Windows may include multiple glazings and spaces to reduce heat transfer.

Low-Transmittance Materials: Low-transmittance glasses and plastics offer a measure of brightness reduction that increases as their transmittance decreases. During daylight hours, the ability to view into a room is reduced. At night, ability to view from a room to the outdoors is reduced. Lower transmittance gives a gloomy appearance to outdoor views.

Directional Transmitting Materials: These include glasses and plastics with prismatic surfaces that are used to obtain the desired directional control of light and luminance. They are used in either horizontal or vertical panels.

Specularly Selective Transmitting Materials: These include the various heat-absorbing and reflecting materials that are designed to pass most visible radiation, but absorb or reflect a portion of the infrared radiation, which would otherwise contribute to cooling loads. Absorbed heat is reradiated indoors and outdoors in approximately equal proportions. Stained glass comes under this classification, as it is selective in the visible portion of the spectrum, though primarily for reasons of appearance rather than as a means of illumination control.

High-Reflectance, Low-Transmittance Materials: Reflective glasses and plastics provide luminance control by having high exterior reflectances. These materials act as one-way mirrors, depending on the ratio of indoor to outdoor illumination. Their low transmittance gives outdoor areas an overcast appearance, even on sunny days. Some selectively admit visible light while reflecting infrared wavelengths that would otherwise add to the cooling load.

Exterior Reflecting Elements: Reflective pavements and similar surfaces increase the amount of ground light entering the building. Reflecting materials or finished roofs below windows have the same effect.

Interior Reflecting Elements: In general, most interior elements should have a high reflectance. This reduces contrast and glare, and is especially important for elements near windows, such as mullions. In particular, ceiling reflectance has a large influence on the amount of daylight (sunlight, skylight, and ground-reflected light) delivered to parts of rooms far from windows and light shelves.

Glazings: High-performance glazings use multiple cavities separated by films and filled with special gases, as well as low-emissivity coatings, to reduce heat transfer.

Electrically Controlled Glazings: Electrically controlled glazings have transmittance properties that are a function of an applied voltage. Currently available products require the application of a low voltage to render them opaque and become clear when the voltage is switched off.

Diffusing Materials: Generally, transmittance and luminance decrease as diffusion increases. The luminance of highly diffusing materials is nearly constant from all viewing angles. Diffusing materials include translucent and surface-coated or patterned glass, plastics, translucent sandwich panels with fiberglass reinforced polymer face, and diffusing glass block.

Shades and Draperies: These include opaque and diffusing shades and draperies for excluding or moderating daylight and sunlight. To darken a room, as for projection, the system must be opaque and completely cover the window.

Trees can be effective shading devices for buildings of low elevation if placed in an appropriate position with respect to windows. Deciduous trees provide protection against glare due to direct sun during the warm months, but transmit sunlight during the winter. Deciduous vines on louvered overhangs or arbors provide similar seasonal shade.

Louvers: Louvers may be fixed or adjustable, horizontal or vertical. They are capable of excluding direct sunlight and reducing radiant heat while reflecting a large portion of sun, sky, and ground light into the interior. In the case of fixed louvers, the spacing and height of the slats should be designed to exclude direct sunlight at normal viewing angles. Overhangs for sun control are often made with louver elements so that light from the rest of the sky can reach the windows.

6.4   Basic Calculations for Daylighting

Continuous time variations of daylight require a determination of total daylight factor to assess daylight availability. The total daylight factor (DF) is composed of SC (sky component), ERC (externally reflected component), and the internally reflected component (IRC). This takes into consideration reflection from interior and exterior surfaces, all calculated individually for each location being considered.

$$DF = SC + ERC + IRC$$

$$ERC = SC \times RF \text{ (reflectance factor of obstruction)}$$

$$SC = \text{Incident skylight} - \text{window losses}$$

IRC is normally calculated using published interreflectance tables.

Illuminance Computation: Illuminance computation involves determination of the sky, sun, and interreflected components of daylight. The sky and sun components are due to light flux that reaches a point directly. The interreflected component results from sunlight and skylight that have initially reached another surface and then been reflected to the point of interest. The lumen method is commonly used to compute illuminances at points in the building or average illuminances over surfaces inside buildings.

The lumen method for calculation of daylighting interior illuminances is similar to the zonal cavity method for electric lighting, and is simple enough to permit manual computation. It provides a simple way to predict interior daylight illumination through skylights and windows. It assumes an empty rectangular room with simple fenestration and shading devices. The basic equation for the illuminance at a prescribed point using the lumen method is the simple formula

$$E_i = E_x(NT)(CU)$$

where

E_i = interior illuminance in lx

E_x = exterior illuminance in lx

NT = net transmittance

CU = coefficient of utilization.

The procedures for determining the net transmittance and the coefficient of utilization differ for toplighting and sidelighting. If both types of systems are being employed, the illuminance can be computed for each and the illuminances added to give the combined effect.

For daylighting systems employing horizontal apertures such as skylights at or slightly above roof level, the *lumen method for toplighting* is used. It is assumed that the skylights are positioned uniformly across the ceiling. (The following three equations were reprinted with the permission of the IESNA, 120 Wall Street, 17th Floor, New York, NY 10005. Taken from the *IESNA Lighting Handbook, 8th Edition.*)

$$E_i = E_{xh} \tau CU \frac{A_s}{A_w}$$

where

E_i = average incident illuminance on the workplane from skylights in lx

E_{xh} = horizontal exterior illuminance on the skylights in lx

A_s = gross projected horizontal area of all the skylights in m^2

A_w = area of the workplane in m^2

τ = net transmittance of the skylights and light well, including losses because of solar control devices and maintenance factors. Refer to Tables 6.1 and 6.2 for data on transmittance for common skylight materials and light loss factors due to dirt depreciation.

CU = coefficient of utilization.

If the window is image preserving, then the illuminances are given by

$$E_i = \tau(E_{xvsky} \, CU_{sky} + E_{xvg} \, CU_g)$$

and for a diffuse window by

$$E_i = 0.5 \, \tau(E_{xvsky} + E_{xvg}) \, (CU_{sky} + CU_g)$$

where

E_i = interior illuminance at a reference point in lx

τ = net transmittance of the window wall

E_{xvsky} = exterior vertical illuminance from the sky on the window in lx

CU_{sky} = coefficient of utilization from the sky

E_{xvg} = exterior vertical illuminance from the ground on the window in lx

CU_g = coefficient of utilization from the ground.

The *daylight factor method* treats the illuminance that occurs at a point inside a room as a fraction of the simultaneous illuminance on an unobstructed horizontal plane outdoors. This ratio is called the daylight factor. Direct sunlight is excluded for both interior and exterior values of illumination.

The advantage of the daylight factor as an indicator of daylighting performance is that it expresses the efficiency of a room and its fenestration as a natural lighting system. The use and limitations of the daylight factor are related to its variability over time when determined from measured values (even under overcast skies), its variability under clear and partly cloudy skies due to the constantly changing sky luminance distribution, its failure to correlate with human

Table 6.1 Typical Light Loss Factors for Daylighting Design (Reprinted with permission from IESNA Lighting Handbook, 8th edition.)

Locations	Light Loss Factor Glazing Position		
	Vertical	Sloped	Horizontal
Clean areas	0.9	0.8	0.7
Industrial areas	0.8	0.7	0.6
Very dirty areas	0.7	0.6	0.5

Table 6.2 Net Transmittance Data of Glass and Plastic Materials (Reprinted with permission from IESNA Lighting Handbook, 8th edition.)

Material	Transmittance (percent)
Polished plate/float glass	80–90
Sheet glass	85–91
Heat-absorbing plate glass	70–80
Heat-absorbing sheet glass	70–85
Tinted polished plate	40–50
Figure glass	70–90
Corrugated glass	80–85
Glass block	60–80
Clear plastic sheet	80–92
Tinted plastic sheet	42–90
Colorless patterned plastic	80–90
White translucent plastic	10–80
Glass fiber reinforced plastic	5–80
Translucent sandwich panels	2–67
Double-glazed clear glass	77
Tinted plus clear	37–45
Reflective glass[a]	5–60

[a]Includes single-glass, double-glazed units, and laminated assemblies. Consult manufacturer's material for specific values.

assessment of the general brightness of spaces, and practical differences in its determination by the field measurement of illuminances.

6.5 Design Considerations for Daylighting

For a minimum investigation at the conceptual stage, daylighting should be evaluated with solar altitudes corresponding to the solstices, around December 21 and June 21, and the equinoxes, around March 21 and September 22. A range of sky conditions should be tested, including, at least, the extremes for a particular orientation (such as solar noon and the earliest and darkest hours when daylighting occurs for a south-facing facade).

It is important that designs be checked for critical conditions, when sunlight may enter spaces. For instance, at northern latitudes, direct sunlight may strike the north-facing facades of buildings on summer evenings. In the winter, the sun will be low in the sky, resulting in deep penetration of shading systems and sidelit spaces. At southern latitudes, the high summer sun can more readily enter spaces through skylights.

The provision of apertures for daylighting must be considered in conjunction with decisions regarding building economics, architectural composition, thermal control, ventilation, acoustics, and other design factors. Serious glare problems must be identified and addressed early in the design process, because fenestration must often be modified to correct problems created by sunlight penetration. Even if windows and skylights are provided for amenity lighting only, daylighting effects must be considered in order to *avoid glare and thermal problems.* It is necessary to consider these two parameters together. The sum of the individual glare source contributions is converted to a criterion called "visual comfort probability," or VCP, which is defined as the percentage of normal-vision observers who will be *comfortable* in that specific visual environment.

Because of the intensity of direct sunlight, direct and indirect glare can be a serious problem. Designs should be developed such that direct sunlight is, or can be, excluded from critical task areas. Sources of reflection should also be identified and eliminated or their reflectance reduced insofar as is possible. Glare produced by reflected sunlight, even for only a few minutes per day, may lead users to take measures that reduce the admission of light for a long time.

Glare increases substantially with larger views of the upper portion of the sky. This can be avoided by limiting the height of the view window head in critical tasks areas, by screening upper window areas from view, or by placing daylighting apertures high enough to be out of the normal field of vision. The task may also be arranged so that the user does not face the window, although this may conflict with user view preferences. Glare can also be reduced by using light colors on interior surfaces. This is critical for areas adjacent to windows, which are typically the surfaces of highest luminance.

Computer Models: A number of general-purpose three-dimensional modeling programs for architectural design are now available that allow the user to model complex spaces and determine patterns of sunlight penetration through apertures in the envelope. Most of these do not calculate illuminances, but that is not required at this stage of design. Only measurement of illuminance is required. Because of the short-term variability of daylight, it is usually necessary to use sensors connected to a data logger with recording capability. An alternative is the use of artificial skies and suns, although only a few such facilities exist, mostly at universities or research institutions. Artificial skies and suns also have limitations.

Physical Model: Luminance-based measurements using calibrated video cameras can record thousands of luminance measurements in a physical model, allowing evaluation of lighting attributes such as luminance distribution, color, and visual performance. These video-based systems can also assist the designer in recording subjective evaluations of the model.

It is generally necessary to make measurements of the total illuminance on an unobstructed horizontal surface (the global illuminance) simultaneously with interior measurements as a record of daylight availability. Other important measures are the diffuse illuminance on an unobstructed horizontal surface and the zenith luminance. The most basic means of determining the diffuse illuminance is screening a sensor with a shadow band and using a correction factor to compensate for the diffuse daylight obstructed by the shadow band. It may also be useful to record daylight on vertical planes of interest, usually facing one or more of the cardinal directions, which necessitates the use of screening devices to cut off ground-reflected light.

6.6   Integrated Daylighting System Control

Daylighting will provide energy savings primarily through reductions in the operation of electric lighting.

Daylight Compensation: If daylighting can be used to replace some electric lighting near fenestration during substantial periods of the day, lighting in those areas should be circuited so that it may be controlled automatically by switching or dimming. For this purpose, photoelectric control is an alternative to manual switching of lights.

Control Compatibility: If a control system is used, check compatibility with the control devices. Install multicircuit switching or preset dimming controls to provide flexibility when spaces are used for multiple purposes and require different ranges of illuminance for various activities. Use occupant/motion sensors for unpredictable patterns of occupancy.

It is important that building operators and occupants be educated as to the purpose and function of daylighting controls if they are to be used. Automatic controls may be deactivated by persons who fail to understand their purpose and function.

Space utilization: Use daylighting in transition zones, in lounge and recreational areas, and for functions where the variation in color, intensity, and direction may be desirable. Consider applications where daylight can be utilized as ambient lighting, supplemented by local task lights. Trim trees and bushes that may be obstructing and creating unwanted shadows.

Large control zones may not adequately respond to different lighting conditions within them. The subdivision of the building into sufficiently small control zones is an important daylighting control consideration.

7

Earth, Lakes, and Building Systems

7.1 Geothermal Energy

What is geothermal energy? Ten miles beneath the earth's surface, an almost limitless source of energy waits to be mined. It's not oil, coal, or natural gas, but the energy from the rocky, hot interior of 420 °C, or 800 °F, or more. This intense heat, known as geothermal energy, turns underground water into the steam that erupts from geysers. These natural pockets of steam, within a mile of the surface, are too rare to provide large amounts of natural geothermal energy worldwide.

Through plate motion and vulcanism, some of this energy is concentrated at high temperature near the surface of the earth. Energy is also transferred from the deeper parts of the crust to the earth's surface by conduction and by convection in regions where geological conditions and the presence of water permit. Geothermal energy as applied to building systems is the thermal energy within the earth's crust. Calculations show that the earth, originating from a completely molten state, would have cooled and become completely solid many thousands of years ago without an energy input in addition to that of the sun. It is believed that the ultimate source of geothermal energy is radioactive decay within the earth.

Because of variation in volcanic activity, radioactive decay, rock conductivities, and fluid circulation, different regions have different heat flows (through the crust to the surface), as well as different temperatures at a particular depth. The normal increase of temperature with depth (i.e., the normal geothermal gradient) is about 13.7 °F per 1000 feet of depth, with gradients of about 5 to 27 °F per 1000 feet being common. The areas that have higher temperature gradients and/or higher-than-average heat flow rates constitute the most interesting and viable economic resources. However, areas with normal gradients may be valuable resources if certain geological features are present.

7.1.1   Introduction to Geothermal Systems

Geothermal systems that produce essentially dry steam are vapor dominated. While these systems are valuable resources, they are rare. Hot water (fluid-dominated) systems are used more common than vapor-dominated systems and can be produced either as hot water or as a two-phase mixture of steam and hot water, depending on the pressure maintained on the production system. If the pressure in the production casing or in the formation around the casing is reduced below the saturation pressure at that temperature, some of the fluid will flash, and a two-phase fluid will result. If the pressure is maintained above the saturation pressure, the fluid will remain single-phase. In fluid-dominated systems, both dissolved gases and dissolved solids are significant. The quality of the fluid varies from site to site, from a few hundred parts per million (ppm) to over 300,000 ppm dissolved solids.

The following characteristics have a major influence on selection of a geothermal system. Because the costs of geothermal systems are primarily front-end capital costs, annual operating costs are relatively low.

- Depth of resource
- Distance between resource location and application site
- Well flow rate
- Resource temperature
- Temperature drop
- Load size
- Load factor
- Composition of fluid
- Ease of disposal
- Resource life.

7.1.2   Applications in Building Systems

The primary applications for the direct use of geothermal energy are space heating, sanitary water heating, and space cooling (using the absorption process). While geothermal space and sanitary water heating are widespread, space cooling is rare. Where space heating is accomplished, sanitary water heating or at least preheating is almost universally accomplished. Domestic water heating in a district space heating system is beneficial because it increases the overall size of the energy load, the energy demand density, and the load factor.

Whenever possible, the domestic hot water load should be placed in series with the space heating load to reduce system flow rates. In many cases, buildings that derive their heat from geothermal fluids operate their heating systems at less than conventional supply water temperatures because of low resource temperatures and the use of heat exchangers to isolate the fluids from the building loop. See Figure 7.1 for a central plant groundwater system arrangement with heat exchangers. In geothermal systems it is frequently advisable to design geothermal heat exchange surface larger than normal surface requirement. The selection of equipment to accommodate these considerations can enhance the feasibility of using the geothermal source.

7.1.3 Components of Geothermal Systems

The primary equipment used in geothermal systems includes pumps, heat exchangers, and piping. While some aspects of these components are unique to geothermal applications, many of them are of routine design. However, the great variability and general aggressiveness of the geothermal fluid necessitate limiting corrosion and scale buildup rather than relying on system cleanup. Corrosion and scaling can be limited through (1) proper system and equipment design or (2) treatment of the geothermal fluid, which is generally precluded by cost and environmental regulations relating to disposal.

The terminal equipment used in geothermal systems is the same as that used in nongeothermal heating systems. However, certain types of equipment are better suited to geothermal design than others. Finned coil, forced-air systems are generally the most capable of functioning under the low-temperature/high-temperature difference conditions. One or two additional rows of coil depth compensate for lower supply water temperatures. While increased delta-T affects coil circuiting, it improves controllability. This type of system should be capable of using supply water temperatures as low as 100 to 120 °F (37.7–48.8 °C).

Radiant floor panel systems are able to use very low water temperatures, particularly in industrial applications with little or no floor covering. The availability of new nonmetallic piping has renewed the popularity of this type of system. In industrial settings, with a bare floor and relatively low space temperature requirements, average water temperatures could conceivably be as low as 95 °F (35 °C).

Heat pump systems take advantage of the lowest temperature geothermal resources. Loop heat pump systems operate with water temperatures in the 25 to 100 °F range, and central station heat pump plants (supplying four-pipe systems) in the 45 °F range. Baseboard convectors and similar equipment are the least capable of operating at low supply water temperature. At 150 °F (65.5 °C) average water temperatures, derating factors for this type of equipment are on the order of 0.43 to 0.45. As a result, the quantity of equipment required to meet the design load is generally uneconomical. This type of equipment can be operated at low temperatures from the geothermal source to provide base-load heating capacity. Peak load can be supplied by a conventional boiler.

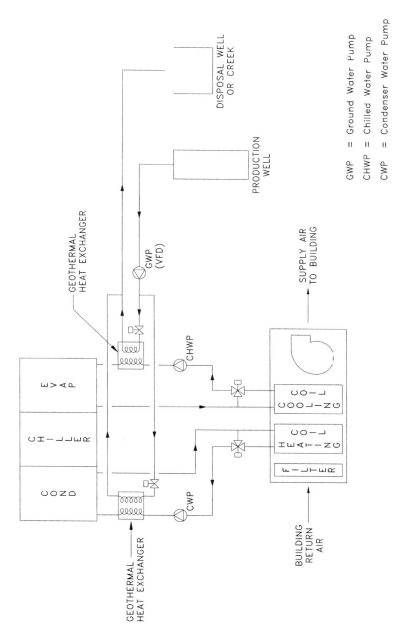

Figure 7.1 Central plant groundwater/geothermal system.

GWP = Ground Water Pump

CHWP = Chilled Water Pump

CWP = Condenser Water Pump

7.2   Thermocline Lakes

Lakes are stratified bodies of water in which temperature declines with the depth. Heat is transferred to lakes by three primary modes: radiation from the sun, convective heat transfer from the surrounding air (when the air temperature is greater than the water temperature), and conduction from the ground.

7.2.1   Introduction

Heating of lakes is primarily accomplished by solar radiation, which can exceed 300 Btu/h per square foot of lake area, but it occurs primarily in the upper portion of the lake unless the lake is very clear. About 40% of the solar radiation is absorbed at the surface. Convective heat transfer to the lake occurs when the lake surface temperature is lower than the air temperature. Wind speed increases the rate at which heat is transferred to the lake, but maximum heat gain by convection is usually only 10% to 20% of maximum solar heat gain. The conduction gain from the ground is even less than convection gain.

Likewise, cooling of lakes is accomplished primarily by back-radiation, which typically occurs at night when the sky is cool, triggering evaporative heat transfer at the surface. Convective cooling or heating in warmer months contributes only a small percentage of the total because of the relatively small temperature differences between the air and the lake surface temperature. The last major mode of heat transfer, conduction to the ground, does not play a major role in lake cooling.

In the winter, the coldest water is at the surface. It tends to remain at the surface and freeze. The bottom of a deep lake stays 5 to 10 °F warmer than the surface. This condition is referred to as winter stagnation. The maximum density of water occurs at 39.2 °F (4 °C), not at the freezing point of 32 °F (0 °C). As spring approaches, surface water warms until the temperature approaches the maximum density point of 39.2 °F. Later in the spring as the water temperatures rise above 45 °F (7.2 °C), the winter lake stratification becomes unstable, and circulation loops begin to develop from top to bottom. This condition is called spring overturn. This pattern continues throughout the summer. The upper portion of the lake remains relatively warm, with evaporation cooling the lake and solar radiation warming it.

The lower portion (hypolimnion) of the lake remains cold because most radiation is absorbed in the upper zone, circulation loops do not penetrate to the lower zone, and conduction to the ground is quite small. The result is that in deeper lakes with small or medium inflows, the upper zone is 70 to 90 °F (32.2 °C), the lower zone is 40 to 55 °F, and the intermediate zone (thermocline) has a sharp change in temperature within a small change in depth. This condition is referred to as summer stagnation.

As fall begins, the water surface begins to cool through back radiation and evaporation. With the approach of winter, the upper portion begins to cool toward the freezing point, and the lower levels approach the maximum density temperature of 39.2 °F. Many lakes do exhibit near-ideal temperature profiles. However, a variety of circumstances can disrupt the profile. These circumstances include (1) high inflow/outflow rates, (2) insufficient depth for stratification, (3) level fluctuations, (4) wind, and (5) lack of enough cold weather to establish sufficient amounts of cold water for summer stratification.

7.2.2 Application and Components in Building Systems

Thermal stratification of water often causes large quantities of cold water to remain undisturbed near the bottom of deep lakes. This water is cold enough to adequately cool buildings by simply being circulated through heat exchangers. A heat pump is not needed for cooling, and energy use is substantially reduced. Heating can be provided by a separate source or with heat pumps in the heating mode. Prechilling or supplemental total cooling are also permitted when water temperatures are between 50 and 60 °F.

The heat pumps are used to transfer heat to or from the air in the building. To put these heat transfer rates in perspective, consider a 1 acre (43,560 ft²) lake that is used in connection with a 10-ton (120,000 Btu/h) heat pump. In the cooling mode, the unit will reject approximately 150,000 Btu/h (44 kw) to the lake. This is 3.4 Btu/h-ft², or approximately 1% of the maximum heat gain from solar radi-

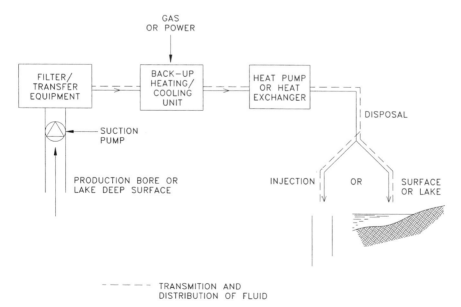

Figure 7.2 Basic geothermal/thermocline lake direct system.

ation in the summer. In the winter, a 10-ton heat pump would absorb only about 90,000 Btu/h (26.3 kw), or 2.1 Btu/h-ft^2, from the lake.

In a closed-loop system, water-to-air heat pumps are linked to a submerged coil. Heat is exchanged to (cooling mode) or from (heating mode) the lake by the fluid (usually a water–antifreeze mixture) circulating inside the coil. In an open-loop system, water is pumped from the lake through a heat exchanger and returned to the lake some distance from the point at which it was removed. The pump can be located either slightly above or submerged below the lake water level. For heat pump operation in the heating mode, this type of system is restricted to warmer climates; water temperatures must remain above 42 °F. See Figure 7.2 for a basic direct use lake system.

8

Energy Storage

8.1   Introduction

A general principle applicable to all physical systems in which mass is neither created nor destroyed is the *law of conservation of energy,* which states that energy is neither created nor destroyed; it is merely changed in form. This principle, together with the laws of electric and magnetic fields, thermodynamics, electric and hydraulic circuits, and Newtonian mechanics, is a convenient means for finding the characteristic relationships of electromechanical energy coupling. Energy required to operate building systems can be stored in thermal storage, electrochemical, passive electric elements, and coupling fields.

Irreversible conversion of electric energy to heat arises from three causes: Part of the electrical energy is converted directly to heat in the resistances of the current paths, part of the mechanical energy developed within the device is absorbed in friction and windage and converted to heat, and part of the energy absorbed by the coupling field is converted to heat in magnetic core loss (for magnetic coupling) or dielectric loss (for electric coupling). If the energy losses in the electrical system, the mechanical system, and the coupling field are grouped, the energy balance may be written in the following form:

Electrical energy input minus resistance losses
= mechanical energy output plus friction and windage losses
+ increase in energy stored in coupling field plus associated losses

Basic property and characteristics desired in a Building System Energy Storage mechanism is that it should be convertible and storage system charging and discharging should be part of base system configuration.

8.2 Thermal Storage

Thermal storage systems remove heat from or add heat to a storage medium for use at another time. Thermal storage for building system applications can involve storage at various temperatures associated with heating or cooling processes. High-temperature storage is typically associated with solar energy or high-temperature heating processes, and cool storage with air-conditioning and refrigeration. Energy may be charged, stored, and discharged daily, weekly, annually, or in seasonal or rapid batch process cycles.

Thermal storage may be an economically attractive approach to meeting heating or cooling loads in the building if one or more of the following conditions apply:

- Loads are of short duration.
- Loads occur infrequently.
- Loads are cyclical in nature.
- Loads are not well matched to the availability of the energy source.
- Energy costs are time dependent (e.g., time-of-use energy rates or demand charges for peak energy consumption).
- Utility rebates, tax credits, or other economic incentives are provided for the use of load-shifting equipment.
- Energy supply from the utility is limited, thus preventing the use of full-size nonstorage systems.

Following is a list of some benefits of thermal storage:

Reduced equipment size

Capital cost savings

Energy savings

Improved system operation.

8.2.1 Applications and Requirements

Thermal storage can take many forms to suit a variety of applications. Some groups of thermal storage applications are: off-peak air conditioning, industrial/process cooling, off-peak heating, and other applications. Whether for heat storage or cool storage, and whether for storing sensible or latent heat, storage designs follow one of two control strategies: full storage or partial storage.

Full storage (load-shifting) designs are those that use storage to fully decouple the operation of the heating or cooling generating equipment from the peak heating or cooling load. The peak heating or cooling load is met through the use (i.e., discharging) of storage while the heating or cooling generating equipment is idle. Full storage systems are likely to be economically advantageous only under one or more of the following conditions:

- Spikes in the peak load curve are of short duration.
- Time-of-use energy rates are based on short-duration on-peak periods.
- There are short overlaps between peak loads and peak energy periods.
- High cash incentives are offered for using thermal storage.
- High peak demand charges apply.

For example, a school or business in which electrical demand drops dramatically after 5 P.M. in an electric utility territory where on-peak energy and demand charges apply between 1 P.M. and 9 P.M. can economically apply a full cool storage system. The brief 4-h overlap between 5 and 9 P.M. allows the full load to be shifted with a relatively small and cost-effective storage system and without oversizing the chiller equipment.

Partial storage more often provides the best economics, and therefore, represents the majority of thermal storage installations. Although they do not shift as much load (on a design day) as full storage systems, partial storage systems can have lower initial costs, particularly if the design incorporates smaller equipment by using low-temperature water and cold air distribution systems. For many applications, a form of partial storage known as load leveling can be used with minimum capital cost.

A load-leveling system is designed with the heating or cooling equipment sized to operate continuously at or near its full capacity to meet design-day loads; thus, equipment of the minimum capacity (and minimum cost) can be used. During operation at less than peak design loads, partial storage designs may function as full storage systems. For example, a system designed as a load-leveling partial storage system for space heating at winter design temperatures may function as a full shift system on mild spring or autumn days.

Storage Media: A wide range of materials can be used as storage media. Desirable characteristics include the following:

- Commonly available
- Low cost
- Environmentally benign
- Nonflammable
- Nonexplosive
- Nontoxic

- Compatible with common building system materials
- Noncorrosive
- Inert
- Well-documented physical properties
- High density
- High specific heat (for sensible heat storage)
- High heat of fusion (for latent heat storage)
- High heat transfer characteristics
- Storage at ambient pressure
- Characteristics unchanged over long use.

8.2.2   Sensible Heat Storage

Common storage media for sensible heat storage in buildings include water, soil, rock, brick, ceramics, concrete, and various portions of the building structure being heated or cooled. In HVAC applications such as air conditioning, space heating, and water heating, water is often the chosen thermal storage medium, providing virtually all of the desirable characteristics when kept between its freezing and boiling points. In lower temperature applications, aqueous secondary coolants (typically glycol solutions) are often used as the heat transfer medium, enabling certain storage media to be used below their freezing or phase-change points.

Sensible thermal energy can also be stored in electrically charged, thermally discharged storage devices. For devices that use a solid mass as the storage medium, equipment size is typically specified by the nominal power rating (to the nearest kilowatt) of the internal heating elements, and the nominal storage capacity is taken as the amount of energy supplied (to the nearest kilowatt-hour) during an 8-hour charge period. For example, a 10-kW heater would have a nominal storage capacity of 80 kWh. Some commonly used electrical storage equipment includes room storage heaters, central storage air heaters, pressurized water storage heaters, and underfloor heat storage.

For high-temperature heat storage, the storage medium is often rock, brick, or ceramic materials for residential or small commercial applications. Use of the building structure itself as passive thermal storage offers advantages under some circumstances. Building elements such as floor slabs and internal walls form the storage, and the materials (concrete, brick, etc.) are the storage media.

The objective for building mass thermal storage is not to maintain a constant temperature in the conditioned space, but rather to limit the rise in temperature over a normal working day. In air-conditioned buildings, running the systems overnight to cool the building mass limits the peak and total cooling loads the following day. The method usually takes advantage of "free cooling" with outside air, which is available during the night in most climates, but the free cooling can be supplemented or replaced by other means of cooling.

8.2.3   Water Thermal Storage

Water is commonly used in both heat storage and cool storage applications. Water has the highest specific heat (Btu/lb.°F) of all common materials and is well suited for thermal storage. The amount of energy stored in a chilled water storage is directly related to the volume of water in the storage times the temperature differential between the entering and leaving water. For example, chillers that cool water in storage to 40 °F (4.4 °C) and coils that return water to storage at 60 °F (15.5 °C) provide a range of 20 °F. The cost of extra coil surface required to provide this range can be offset by savings in pipe size, insulation, and pumping energy. Further, fan energy costs are reduced if the extra coil surface is added as face area rather than as extra rows of coils.

Water Storage Enclosure: A perfect storage tank would deliver water at the same temperature at which it was stored. It would also require that the water returning to storage neither mix nor exchange heat with the stored water or the tank. In practice, however, both types of heat exchange occur. Water storage is based on maintaining a state of thermal separation between cool water and warm water.

Tanks are available in many shapes; however, vertical cylinders are the most common. Tanks can be located above ground, partially buried, or completely buried. Tanks are usually at atmospheric pressure, unpressurized, and may have clear-span spherical dome roofs or column-supported flat roofs. Exposed tank surfaces should be insulated to help maintain the temperature differential in the tank. Insulation is especially important for smaller storage tanks because the ratio of surface area to stored volume is relatively high.

Heat transfer between the stored water and the tank contact surfaces (including divider walls) is a primary source of capacity loss. Not only does the stored fluid lose heat to (or gain heat from) the ambient by conduction through the floor and wall, but heat flows vertically along the tank walls from the warmer to the cooler region. Exterior insulation of the tank walls does not inhibit this heat transfer.

Space Requirements: The location and space required by a thermal storage system are functions of the type of storage and the architecture of the building and site. Building or site constraints often shift the selection from one option to another.

Chilled water systems are associated with large volume. As a result, many stratified chilled water storage systems are located outdoors (such as in industrial plants or suburban campus locations). A tall tank is desirable for stratification, but a buried tank may be required for architectural or zoning reasons. Tanks are traditionally constructed of steel or prestressed concrete. A systems supplier who assumes full responsibility for the complete system performance often constructs the tank at the site and installs the entire distribution system.

8.2.4 Latent Heat Storage

Common storage media for latent heat storage include water-ice, aqueous brine-ice solutions, and other phase-change materials such as hydrated salts and polymers. Water has the highest latent heat of fusion of all common materials—144 Btu/lb (80 cal/gm) at the melting or freezing point of 32 °F (0 °C). Other than water the most commonly used material for latent heat thermal storage is a hydrated salt with a latent heat of fusion of 41 Btu/lb (22.7 cal/gm) at the melting or freezing point of 47 °F (8.3 °C) and a density of 93 lb/ft^3 (1.5 gm/cm^3). Hydrated salts have been in use for many decades, often encapsulated in plastic containers. This material is a mixture of inorganic salts, water, and nucleating and stabilizing agents.

For building systems air-conditioning applications, water-ice is the most common storage medium. The challenge is to find an efficient and economical means of achieving the heat transfer necessary to alternately freeze and thaw the storage medium. Various methods have been developed to limit or deal with the heat transfer approach temperatures associated with freezing and melting; however, leaving fluid temperatures (from storage during melting) must be higher than the freezing point, while entering fluid temperatures (to storage during freezing) must be lower than the freezing point. Ice storage can provide leaving temperatures well below those normally used for comfort and nonstorage air-conditioning applications. However, entering temperatures are also much lower than normal.

8.2.5 Ice Storage

The latent heat of fusion in ice storage systems requires a capacity of about 5.5 ft^3/ton-hour (0.055 m^3/kwhr) for the entire tank assembly, including piping headers and water in the tank. Buildings commonly use following ice storage systems:

Ice-on-Coil Systems (External and Internal Melt): Ice-on-coil systems are available in many configurations with differing space and installation requirements. Because of the wide variety available, these systems often best meet the unique requirements of many types of buildings. Some key performance data requirements of an external melt system are:

- Evaporator and suction temperatures at start of ice build
- Evaporator and suction temperatures at end of ice build
- Ice thickness at end of ice build
- Time to build ice.

From these data, determine actual efficiency. This way, refrigeration capacity deviation from published ratings can indicate refrigerant loss or surface fouling.

In the case of a internal melt-ice-on-coil storage system the following performance data are important:

- Secondary coolant temperature and suction temperature at start
- Secondary coolant temperature and suction temperature at end
- Secondary coolant flow
- Tank water level at start
- Tank water level at end
- Time to build ice.

Based on the above measured flow data and heat balance, determine efficiency at start versus theoretical, and again efficiency at end versus theoretical. Compare this with published ratings.

Ice storage can be integrated as a part of the building structure, by installing bare coils in concrete wall cells. The bare steel coil concept can be used with direct cooling, in which the refrigerant is circulated through the coils, and the water is circulated over the coils to be chilled or frozen. This external melt system has very stringent installation requirements. Coil manufacturers do not normally design or furnish the tank, but they do provide design assistance, which covers distribution and air agitation design as well as side and end clearance requirements. These recommendations must be followed exactly to ensure success.

Encapsulated Ice: This system uses plastic containers filled with deionized water and an ice-nucleating agent. The plastic containers must be flexible to allow for change of shape during ice formation. During discharge as ice melts, the plastic containers return to their original shape.

Ice Harvesting Systems: This system separates ice formation from ice storage. Ice harvesting system can melt the stored ice very quickly. During the ice generation mode, the system is energized if the ice is below the high ice level. Challenge is to accurately quantify remaining ice in the tank. When ice is floating in the tank, the water level will always be constant, so it is impossible to measure ice inventory by measuring water level.

The conductivity of the ice-water solution increases as water freezes, dissolved solids are forced out of the ice into the liquid water, thus increasing their concentration in the water. Accurate ice inventory information can be maintained by measuring conductivity. The following data are important in commissioning an ice-harvesting system:

- Suction temperature at start
- Suction temperature at harvest
- Harvest time/condensing temperature
- Time from start to full signal

- Tank water level at start
- Tank water level at bin full signal.

From the above data determine efficiency and capacity and compare with published ratings.

8.2.6 *System Configurations*

Thermal storage can be configured for various operating modes. Optimum configuration is related to storage technology, discharge temperature needed, and operating mode flexibility required. Common operating modes used in facilities are shown in Figure 8.1.

Open Systems: Chilled water, external melt ice-on-coil, and ice-harvesting systems are all open chilled water piping systems. Due to the potential for drain-down, the open nature of the system, and the fact that the water being pumped may be saturated with air, the design must provide the piping details carefully to prevent pumping, drain-down, or piping problems. Drain-down must be prevented by isolation valves, pressure-sustaining valves, or heat exchangers.

Closed Systems: Closed systems normally circulate a glycol solution coolant either directly to the cooling coils or to a heat exchanger interface to the chilled water system. A domestic water makeup system should not be the automatic makeup to the glycol solution coolant system. An automatic makeup system that pumps a premixed solution into the system is recommended, along with an alarm signal to the building automation system to indicate makeup operation. An accurate estimate of volume is required. The coolant must contain inhibitors to protect the steel and copper found in the piping system. The water should be deionized.

Piping for ice storage systems can be configured in a variety of ways. The chiller may be located upstream or downstream of the building load. See Figure 8.2 for an upstream chiller arrangement.

8.3 Electrochemical Storage

First we consider some basics like *What is chemical energy?* When gasoline mixes with air and ignites, a rapid and violent transformation takes place. This chemical reaction produces carbon dioxide, water, and enough energy to turn a generator shaft. Chemical reactions inside the human body change the compound glucose into carbon dioxide and water. Although these reactions are less violent than the one involving gasoline, they produce enough energy to power the human body. Countless different types of chemical reactions occur throughout the universe, and they all involve changing the composition of matter. This change consists of a regrouping of the atoms in one set of molecules and breaking the complex bonds that hold them together to produce a different arrangement. Some

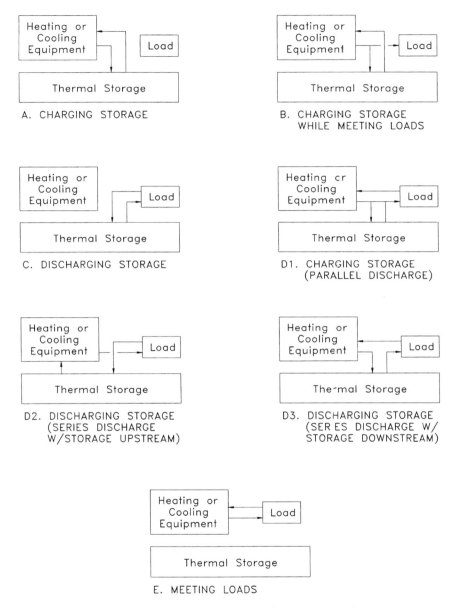

Figure 8.1 Thermal storage operating modes. (Reprinted by permission of the American Society of Heating, Refrigerating and AirConditioning Engineers, Atlanta, Georgia, from the *1995 ASHRAE Handbook—Applications*)

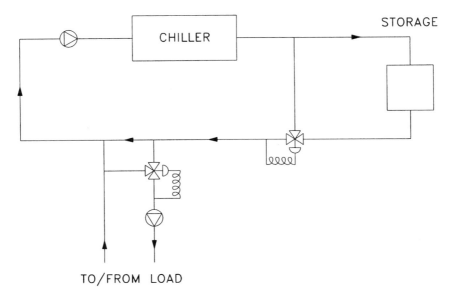

A. CHARGING STORAGE

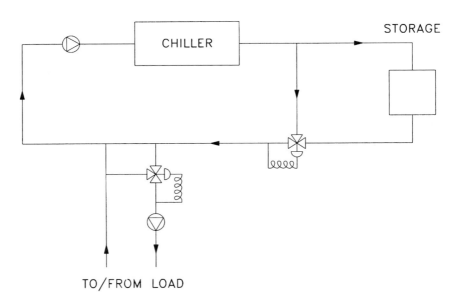

B. DISCHARGING STORAGE

Figure 8.2 Thermal storage system with chiller upstream. (Reprinted with permission of the American Society of Heating, Refrigerating and AirConditioning Engineers, Atlanta, Georgia, from the *1995 ASHRAE Handbook—Applications*)

reactions happen slowly and release only a small amount of energy. Other reactions require a separate source of energy before they can occur. For example, in chemical reactions known as photosynthesis, plants use the energy in sunlight to join together molecules of carbon dioxide and molecules of water to produce glucose and oxygen.

Electrochemical energy is converted into electric energy in batteries and fuel cells. The former are energy storage devices and store the chemical energy within the battery. Fuel cells are energy conversion devices and produce energy all the time. The required chemicals (fuel and oxidant) are supplied in the cell.

8.3.1 Batteries

A battery is an assembly of electrochemical cells that convert the energy produced by chemical reactions into low-voltage, direct current electricity. Similar to other electrochemical cells, each cell contains three major components: a positive electrode or cathode, a negative electrode or anode, and the electrolyte.

Primary Battery: Primary batteries convert chemical energy into electric energy once only, and the original chemical state cannot be regained by passing electric energy through the battery. They are discharged once and are discarded after use.

Secondary Battery: After discharge of a secondary battery, their original chemical state may be regained by passing through the battery a quantity of electric energy equivalent to that drawn during discharge. Secondary batteries are, therefore, distinguished from primary batteries by the feature of rechargeability. Secondary batteries are commonly used in building system storage applications.

Battery capacity to deliver a certain amount of energy is generally provided in terms of energy density. *Energy density* is the energy delivered per unit weight (or unit volume).

$$\text{Energy density} = \text{ampere hour capacity} \times \text{cell voltage}$$

Energy density is usually expressed in Wh/lb (Wh./kg) or Wh./in^3 (Wh./cc). In addition to the energy density of a cell, another important parameter is the rate at which the energy, that is, the power, can be delivered. *Power density* is the power delivered by the battery per unit weight or volume:

$$\text{Power density} = \text{current} \times \text{voltage} / \text{battery weight}$$

Units are W/lb or W/in.^{3}

Batteries rarely use 100% of the theoretical ampere-hour capacity. For primary cells, the reason is that mass transport and resistance effects limit the *utilization* of reactants, whereas for secondary cells, the energy used in one cycle is maintained below 100% in order to maximize battery life. For secondary battery applications, high-energy density is useful only if the battery can be repetitively discharged, that is, long cycle life at high reactant utilizations. Depth of discharge, or utilization, is defined as the ratio of the number of ampere-hours delivered to the theoretical number of ampere-hours calculated. Utilization depends on the battery discharge rate, and higher utilizations are achieved at slower discharge rates (i.e., lower currents).

The ampere-hour capacity available from the battery to a preselected voltage depends on the value of the current. At high load (i.e., high currents), the battery delivers far less capacity than at low load. A *capacity/time quotient* is, therefore, defined to normalize the discharge rate for cells of differing capacity. For example, C/5 is the current that completely discharges (or charges) the battery in 5 hours; 4C is the current that discharges the cell in $1/4$ hour. These values are independent of the battery size and weight, but are characteristic only for the type of battery. For example, a radio battery has a discharge rate of about C/100 while a building system requires a C/3 discharge rate. *Shelf life* is the loss of capacity of the cell during storage. The period beyond which it becomes uneconomical to store the battery is known as its shelf life.

Cycle life is the number of times a secondary battery can be discharged (or charged) before the battery voltage falls below a prescribed value.

Discharge Voltage Characteristics: When a cell is discharged, that is, a current is drawn from the cell, the voltage immediately decreases owing to electrode and ohmic polarization. There is then usually a period during which the cell voltage remains essentially constant. Subsequently, there is a sudden change in voltage with time. This point of inflection indicates that little useful energy can be obtained by continuing discharge beyond this point.

The shape of this curve depends on the battery and the discharge rate. There is a similar, though opposite, curve during charge, at which further application of the charge current results only in gas evolution at the electrodes rather than re-forming the chemical reactants.

8.3.2 Battery Design and Types

A large number of primary and secondary batteries are available or are being developed to satisfy the building systems. There is no one universal battery that can satisfy all these requirements and each has its particular market niche. See Table 8.1 for types of commercial batteries applied in building systems.

Battery Design Features: As with other electrochemical cells, care is taken in the design of the battery to minimize cell polarization. This is achieved in several ways. First, the electrodes are fabricated to maximize their surface area to volume ratio; this is achieved by using porous electrodes. Thus, a high specific area is formed at the interface between the active materials of the anodes and cathodes and of the electrolyte phase in order to obtain high currents during charge and discharge. At the same time, good electronic conduction must be achieved

Table 8.1   Commercial Battery Types

	Commercial Primary Batteries		
Type	Cathode	Anode	Electrolyte
LeClanche	MnO_2	Zn	NH, Cl/$ZnCl_2$
Alkaline manganese	MnO_2	Zn	KOH
Rubens (or Duracell)	HgO	Zn	KOH
Panasonic high energy	$(Cf_x)_n$	Li	Propylene carbonate with $LiClO_4$
Lithium (GTE)	$SOCl_2$	Li	$SOCl_2$ with LiCi–$AlCl_3$
Lithium (Mallory)	$SOCl_2$	Li	Propylene carbonate with $LiClO_4$
Lithium (Saft)	CuO	Li	Propylene carbonate with $LiClO_4$
	Secondary Batteries		
Type	Cathode	Anode	Electrolyte
Lead acid	PbO_2	Pb	H_2SO_4
NiCad	NiOOH	Cd	KOH
Nickel–iron	NiOOH	Fe	KOH
Silver–iron	Ag_2O_2	Zn	KOH
Zinc–halogen	Br_2 (or CI_2)	Zn	$ZnBr_2$ (or $ZnCi_2$)
Zinc–air	O_2	Zn	KOH
Nickel–hydrogen	NiOOH	H_2	KOH

between the active battery material and the current collector which, in turn, must be stable and noncorrodible. Several types of electrode structures have been developed that satisfy this condition depending on the nature of the battery.

Ohmic polarization is reduced by using electrolyte concentrations corresponding to maximum specific conductivity and by minimizing the anode-to-cathode distance. In order to have small anode-to-cathode separations, care must be taken that the electrodes are not in contact. This is particularly necessary for secondary batteries where electrode shape changes occur during cycling with a resultant change in this distance. As a result, a separator is placed in the electrolyte between the two electrodes. This material is selected for maximum stability in the electrolyte between the two electrodes. This material is selected for maximum stability in the electrolyte, low cost, and optimum porosity to minimize concentration gradients (and, hence, concentration polarization) in the electrolyte.

The remainder of the design considerations involve electrode configurations, low cost, stable case materials, methods of sealing, and low-cost manufacturing techniques, all of which are specific for a particular battery system. However, as a result of added materials to achieve a practical battery, practical energy densities are only about one quarter to one fifth of the theoretical value based on the weight of electroactive materials and electrolyte.

Battery design criteria are related to cost, performance, durability, safety, and environmental considerations. Hence, there is the requirement for high cell voltages, reversible electrode reactions, high electrolyte conductivities, and minimal volumetric differences between reactants and products. Operating costs are influenced by cycle life, maintenance requirements, and energy efficiency of the charge–discharge cycle.

8.3.3 Fuel Cells

A fuel cell is an electrochemical device that continuously converts the chemical energy of the fuel and the oxidant directly to electric energy without an intermediate combustion process. Unlike a battery, a fuel cell does not run down or require recharging. It will continue to operate as long as the fuel (hydrogen) is fed to the anode and the oxidant (air) is fed to the cathode. The electrodes, which are solid material of high internal area, act as reaction sites where the electrochemical transformation of the fuel and oxidant occurs. These reactions produce electrons that flow from one electrode to the other when the electrodes are connected together through an external circuit. Several types of electrolyte may be used; the particular type is determined by the type of fuel and oxidant and by the cell operating temperature. The electrode reactions are

$$2\,H_2 \rightarrow 4\,H^+ + 4\,e \qquad \text{(anode)}$$

$$O_2 + 4\,H^+ + 4\,e \rightarrow 2\,H_2O \qquad \text{(cathode)}$$

giving as the overall cell reaction

$$2\,H_2 + O_2 \rightarrow 2\,H_2O$$

Fuel Cell Systems: A practical fuel cell power system is composed of three major subsystems: fuel processor, fuel cell, and dc-to-ac inverter. Fuel, which may be basically any fossil fuel (oil, natural gas), is fed into a fuel processor. The fuel processor converts the fossil fuel into a gas suitable for the fuel cell (i.e., mainly hydrogen); the gas entering the fuel cell also contains other constituents such as carbon monoxide and carbon dioxide depending on the fuel processor system, coal gasifier, steam reformer or direct gas fired. The third subsystem of the fuel cell power plant is the inverter to convert the direct current of the fuel cell into alternating current. See Figure 8.3 for a schematic of such an arrangement.

Several types of fuel cells are being developed. The cells, characterized by the nature of the electrolyte, are alkaline, acid, and molten carbonate fuel cells.

Fuel Cell Characteristics: The fuel cell is an environmentally acceptable device since it is quiet and has no moving parts. Because it is not a combustion device, emissions such as NO_x, CO, and unburned hydrocarbons are not a problem. Such low-level emissions, coupled with the fuel cell's quiet, water-conserving operation, result in environmental acceptability and siting flexibility.

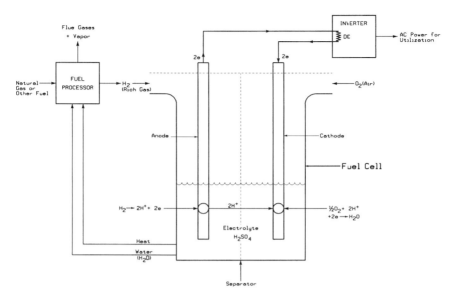

Figure 8.3 Schematic of fuel cell power generation.

A single fuel cell normally generates power at approximately 0.5 to 1 V and can be connected in series stacked with other cells to obtain almost any desired voltage. The current produced is a function of the area of the single cells. The range of sizes, the modularity, and the load-following capabilities make the fuel cell system an attractive candidate for power generation in a variety of applications, including on-site and central plants for commercial, industrial, and residential use.

The major advantage of fuel cells is that they are highly efficient energy conversion devices. In addition, because the system is located at the point of use, the waste heat generated by the irreversibilities of the electrode reactions is recoverable, thus further improving the system efficiency.

8.3.4 *Fuel Cell and System Efficiency*

The efficiency of converting gaseous fuels to dc power in a fuel cell power plant is a function of thermodynamic voltage and current efficiencies, and of the heating value of the composition of the fuel which may contain nonelectrochemically active, combustible species.

Thus the total fuel cell efficiency (E_{fc}) for the conversion of chemical energy in the feed gas to dc power is then defined as follows:

$$E_{fc} = E_T \times E_v \times E_I \times E_H$$

where

E_T = thermodynamic efficiency

E_I = current efficiency

E_v = voltage efficiency

E_H = heating value efficiency.

Fuel Cell System Efficiency: The previous equation describes only the fuel cell power section of the fuel cell system; the fuel processor, power conditioner, and waste heat utilization are not included. The overall efficiency for a fuel cell system (E_s) without waste heat utilization is given by

$$E_s = \frac{\text{ac power}}{\text{HHV of raw fuel into fuel processor}}$$

$$= E_{fp} \times E_{fc} \times E_{pc}$$

where fuel processor efficiency

$$E_{fp} = \frac{\text{LHV gaseous fuel from fuel processor to fuel cell}}{\text{HHV raw fuel into fuel processor}}$$

and power conditioner efficiency

$$E_{pc} = \frac{\text{ac power}}{\text{dc power}}$$

Additional energy can be recovered in the system for heating or by integrating to produce additional electricity. For a system employing waste heat utilization, the fuel cell system efficiency (E_{SE}) is given as

$$E_{SE} = \frac{\text{fuel cell} + \text{electricity produced} + \text{cogeneration BTUs}}{\text{HHV raw fuel into fuel processor}}$$

8.4   Passive Electrical Power Storage

Passive electrical storage devices store energy in the coupling of the device to the power system. The coupling fields are electric and magnetic fields. These fields are physical phenomena: invisible lines of force that occur whenever electrical power is used.

1. *Electric and Magnetic Fields:* Electric and magnetic fields are decoupled below radiofrequencies and are considered separate fields at low frequencies.

2. *Electromagnetic Fields:* Electric and magnetic fields at radiofrequencies and above are coupled together and referred to as electromagnetic fields.

This passive storage device operates low frequencies, and thus the separate terms electric, or magnetic fields will be used.

An electric field represents a force that electric charges exert on other charges. An electric field begins on a charge and ends on a charge, although we picture lines of force emanating from the source like quills on an excited/alarmed porcupine. The electric field is a function of *only* the electric system voltage level. The higher the voltage, the stronger the electric field will be. The unit of measurement of the electric field is volts per meter. The conductor does not have to have current flowing to generate an electric field. The conductor has to be energized to have an electric field present.

A magnetic field is a physical phenomenon of invisible lines of force. Since a magnetic field is a function of moving charges (current), the magnetic field can be looked on as closed loops of force lines such as ripples from a pebble dropping in the middle of a pond. With alternating current, the ripples would reverse with each change in current flow. The magnetic field is a direct function of the amount of electric current flowing in a conductor. The larger the amount of current, the stronger the magnetic field will be. A popular model is that of the self-exciting dynamo: A conductor rotating in a magnetic field creates an electron flow. If the flow passes through a coil, the coil itself creates a magnetic field that keeps the flow moving as long as the conductor keeps rotating. The components described in the following sections are basic passive power storage devices.

8.4.1 Inductors

An inductor is an electrical circuit element whose behavior is described by the fact that it stores electromagnetic energy in its magnetic field. This feature gives it many interesting and valuable characteristics. In electromagnetic radiation, electric field strength is considered the driving force and magnetic field strength the response. In mechanical systems mechanical force is always considered as a driving force and velocity as a response. Four basic points need to be remembered in order to understand electromagnetic interaction. The first point is that a charged particle moving with *uniform* motions creates around itself a magnetic field; the second point is that a magnetic field forces a change of direction in a charged particle moving at an angle through it; the third point is that a charged particle moving with *accelerated* motion produces an electromagnetic wave; the fourth point is that an electromagnetic wave accelerates a charged particle. In its most elementary form, an inductor is formed by winding a coil of wire—often copper—around a form that may or may not contain ferromagnetic materials. Energy storage in inductors is

$$W = \tfrac{1}{2}LI^2.$$

where

L = inductance

I = current through the coil of wire

W = energy stored in the magnetic field.

The unit of inductance is called the henry (H), in honor of the American physicist Joseph Henry.

8.4.2 *Capacitors*

A capacitor is an electrical circuit element that is described through its principal function, which is to store electric energy. This property is called *capacitance.* Capacitance is that property of a system of conductors and dielectrics that permits the storage of electrically separated charges when a potential difference exists between the conductors. In its simplest form, a capacitor is built with two conducting plates separated by a dielectric.

Energy storage in capacitor is

$$W = \tfrac{1}{2}CV_p^2$$

where

W = energy stored in the capacitor

C = capacitance in farads

V_p = peak voltage across the capacitor terminals.

Because the amount of the stored energy varies with the square of the peak voltage, the capacitor's ability to sustain load power during an outage drops off twice as fast as reductions in line voltage. This is why sometimes few cycle outage does not shut down equipment.

8.5   Active Electrical Power Storage (Solar Cells):

Photovoltaics (PV) cells convert sunlight into electric power. PV cells and modules have found many commercial applications ranging from solar powered calculators and watches to arrays making a few megawatts of peak power for utility

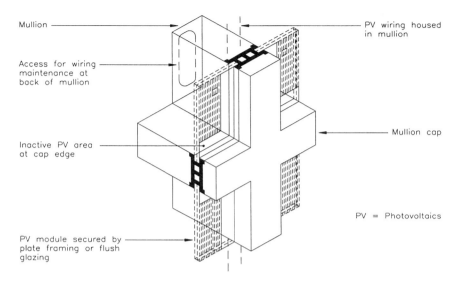

Figure 8.4(a)   Building system integrated, glass based PV glazing

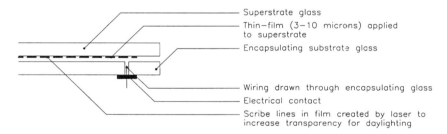

Figure 8.4(b)   Thin-film superstrate-type PV module

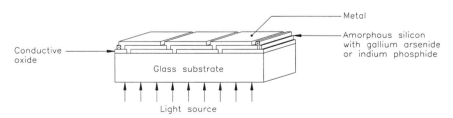

Figure 8.4(c)   Diagram of an Amorphous-silicon module with three series connected cells

generation. The output of solar arrays is direct current; as a result, it must be converted and conditioned if it is intended to feed Power into Building Alternating Current System.

One of the easiest way to incorporate Photovoltaics into Building Systems is to use glass-based PVs for certain wall glazing. See Figure 8–4(a) and 8–4(b). Glass based PVs come in two formats: crystalline silicon or thin-film amorphous silicon. Conversion efficiency range from 12.5% for single-crystal silicon to 7% for polycrystalline silicon to 5% for amorphous silicon. The cost of Photovoltaic and Solar Power range from 8 to 15¢/kwhr with installation cost of $8–$15/Watt.

A typical solar cell consist of either Crystalline or Amorphous Silicon treated with thin layer of cadmium sulfide, gallium arsenide, indium phosphide, or similar compound see Figure 8–4(c). This produces semiconductor Junction between materials. Sunlight impinging on cell creates electron flow hence produces electric direct current flow.

A new technology known as "ac PV modules" integrates a dc-ac "micro-inverter" directly into large-area PV modules thus eliminating the separate inverter used to transform the dc electricity generated by the PV array into ac electricity.

The cost calculation for building integrated PV (BIPV) has many components, including design, hardware (PV modules, power electronics, support structure, and wiring), installation, utility interconnection, metering, maintenance, geographical location, and financing. Systems installed in lower latitudes produce more power than higher latitudes due to greater annual solar resources.

9

Integrated Building Systems Engineering

9.1 Issues and Problems

All systems can be difficult to coordinate because they combine electrical engineering to design the power distribution system, mechanical engineering to provide an understanding of the thermal dynamics, and construction managers to coordinate and install the components of the system. Building M/E system design engineers are poised on a tightrope among rebalancing energy, indoor air quality, power quality, and environmental concerns. Applied integrated technology plays a very important role in examples of current design and construction practices.

In reality there are problems associated with the design of ideal, functional, reliable, and building-worthy systems. As an engineer you must always battle Murphy's law: "if something can go wrong, it will." Philosophically Murphy's law is an aspect of the law of entropy, but ignorance of the law is no excuse. Some common complaints on building systems deficiencies are:

1. Building system capability is often less than promised and expected.
2. Operation cost is not as promised and overruns occur often.
3. There is no room for maintenance and maintenance procedures are complex and error prone.
4. Documentation is inappropriate and inadequate.
5. Building system design lacks human interaction and interoperability.
6. An individual subsystem cannot be integrated.
7. Equipment does not perform according to specifications.
8. The building system does not accommodate user needs.

9. Carcinogens such as asbestos, fibers, dusts, odors, heat/humidity, smog, and poisonous carbon monoxide—air quality issues.

10. Refrigerant leakage, reducing the ozone layer.

11. Carbon dioxide from fuel burning, causing "greenhouse effect" global warming, with uncertain consequences, probably including the increased melting of polar ice and climate changes.

12. Acoustic noise is irritant. Below a certain level, noise is only a nuisance. Above that level, depending on frequencies, noise can injure hearing. Such noise is most common inside factories and on airports.

13. RF noise, electromagnetic interference.

14. Ground current loops are sources of noise and sensitive equipment malfunction and to top it, these currents cause corrosion.

15. Electric and magnetic fields from power distribution systems and extra-low-frequency radio transmitters (ELF) are sources of health hazards.

16. By the way sanitary is backed up.

Many potential difficulties emanate from these deficiencies. Among these are inconsistent, incomplete, or otherwise imperfect identification of user requirements for a building interoperable system. A major objective of building system engineering is to improve effectiveness, including availability, reliability, maintainability, quality, and interoperability and at the same time to reduce first cost and operation cost. A key way to improve building systems performance is to improve design engineering processes.

9.2   Integration Issue or Problem Solution

To achieve a high measure of system functionality, efficiency, and effectiveness, system design features should include features to allow maintenance, expansion, and modification throughout all phases of system life cycle. This life cycle begins with conceptualization and identification of needs, moves through specification of system components and with architectural structures, to the life cycle analysis of the system, to the construction level design of the system, to system installation and commissioning, and ends with operational implementation and associated maintenance. This system design and commissioning cycle can be called "the systems engineering life cycle."

Integration and organization of a large collection of building components and systems require the following:

1. A review of the relative thermal, ventilation, and lighting role of building envelopes versus their internal heating/cooling equipment. It is in-

correct to artificially separate the power/lighting system from the HVAC system with which it intimately interacts.

2. A thorough understanding of *envelope components,* and internal components, typically used to integrate HVAC and lighting and building load cycles to reduce the need for energy and power.

3. The basics of the process by which mechanical and electrical systems are integrated into building design.

4. Critical loads within the facility should be pinpointed to determine how best to reliably serve them. The reliability of an electric system is only as good as that of its weakest element. Therefore, it may be necessary to provide redundancy at anticipated weak points in the system to establish reliable power paths and furnish individual standby power packages for them.

9.3 Building System Project Constraints

The flowchart (Figure 9.1) demonstrates the interdisciplinary nature of building system designs in general and its particular connection with power, lighting, HVAC, and daylighting (fenestration). This approach, which is most often referred to as the systems design approach, will be discussed.

1. Owner-designer-user group. The owner establishes the cost framework, both initial and operating. A part of both of these may be a rent structure, which in turn determines and is determined by the space usage. If the owner is also the occupant, the cost factors change somewhat but remain in force. The architect determines the architectural nature of the space; many of these data are detailed in the building program. Obviously the architect and building system designer (who may be one person) should interact in this aspect of building design.

2. The jurisdictional authorities may include:

   NEC   National Electrical Code

   FEA   Federal Energy Administration

   GSA   General Services Administration

   NCSBCS   National Conferences of States for Building Codes and Standards

   ASHRAE   American Society of Heating, Refrigerating, and Air Conditioning Engineers

   IES   Illuminating Engineering Society

   NBS   National Bureau of Standards

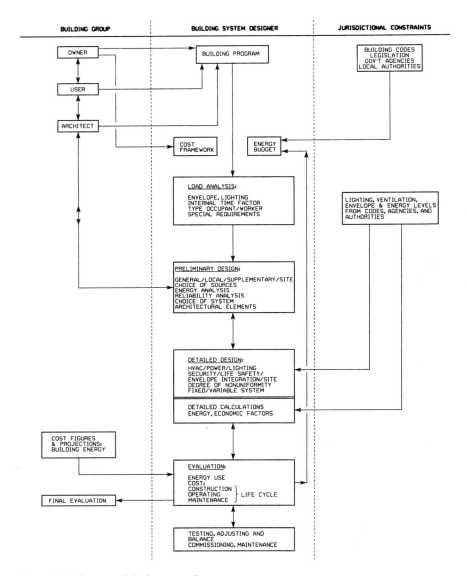

Figure 9.1 Integrated design procedure.

NFPA   National Fire Protection Association

EPA   Environmental Protection Agency

FM   Factory Mutual

IEEE   Institute of Electrical & Electronic Engineers

UBC   Uniform Building Code

BOCA   Building Officials & Code Administrators International, Inc.

UMC   Uniform Mechanical Code

NSPC   National Standard Plumbing Code

OSHA   Occupational Health and Safety Agency

State, county, and municipal codes

It would be a wonderful life if we could work in whatever way we felt would produce the best design, but in the real world, we are subject to a large number of constraints. It helps to recognize what they are and learn either how to overcome them or to live with them.

Specifications and Standards: Before starting to conceptualize a design, learn the specifications and standards that apply. If you do not, and violate a specification, the cost of redesign may be high. The major specification list is long:

MIL specs

Clients general specs.

Foreign specs, including, for example:

    Canadian Standards Association (CSA)

    Verband Deutscher Electrotechniker, E.V. (VDE)

    International Electrotechnical Commission (IEC)

    Deutsche Industria—Norm (DIN)

    International Standards Organization (ISO)

    Your own company and department standards.

The project has a budget and a schedule that you may or may not have influenced. This will constrain the depth of study and experimentation you might like in order to turn out the best possible design. A tight budget and schedule require conservative design which will certainly work but may not be optimum.

9.4   Cost Issues and Constraints

The principal area of involvement is that of first cost, energy budgets, and life cycle cost, all of which affect every aspect of system design including sources type, equipment selection, HVAC system, equipment placement, and even maintenance schedules. For this reason the first step in the system design procedure is to establish the *project system cost framework and the project energy budget.*

Money or cash flow units such as dollar, pound, mark, yen, pence, cents, etc. and their conversions are very much like any other engineering flow unit such as ampere, cubic feet per minute, lbs/h, and gallons/minute. If money outflow increases, the engineering friction/heat increases just like any other flow within fixed passage/channel. Considerations of cost enter into almost all design decisions. A not-for-profit organization such as a university or government laboratory has a budget. It is supposed to produce the best and most value it can for that budget, and it pays the engineer to help it do so. The inevitable tradeoffs between first cost and operating cost cannot intelligently be made unless the cost structure is clearly understood. There are no cookbook formulas. (There are, but most are platitudes). The following guidelines should be of considerable assistance both in avoiding unpleasant surprises when a job is estimated and in preparing cost analyses:

1. Decide at the outset what cost criteria will be applied, that is, the relative importance of first cost, operating costs, annual owning costs, and life-cycle costs. Tradeoff decisions are required between first cost and operating costs.

2. The impact on the operating cost of the entire building system must be studied, and the apportionment of costs determined.

3. With the continued instability in energy source provider, it is suggested that even in an *annual* owning cost comparison, two different energy costs be used and the impact of unreliable source be studied. The only practical means of accomplishing this is by computer program.

4. Design for automation instead of for manual work.

5. Minimize the total cost of the M/E systems. This cost is the sum of:
    *   Purchase price
    *   System operating cost over life (energy, power, operating labor)
    *   System maintenance cost over life
    *   System downtime cost over life

6. Avoid "standard design" as expensive laziness.

7. Material specification: Choose the most cost-efficient. Minimize the net per unit cost of each component, which is the sum of:
    *   Per unit material cost
    *   Per unit vendor services cost
    *   Per unit labor cost, with overhead

8. Study vendor manuals to properly apply products.

9. Finally, pretend that it is your own money.

We are accustomed to pay for one benefit by trading off a reduction in another benefit. There is a body of mathematics that optimizes the mix of benefits. As a designer, you feel that most of these costs are not your problem. They are. An engineer is paid to design a building system that will perform but not cost. At one time, the process of cost reduction after initial design was creatively labeled Value Engineering, and some contractors promoted whole departments to second-guess other engineers. It's less irritating if the designer him- or herself does it before somebody else does.

9.5   Energy Problems and Interpretation of Issues

There are several proposals to establish an upper limit for building energy consumption in operation. In their simplest form, such regulations would specify the maximum allowable energy input per building-area unit per year: British thermal units per square foot per year (Btu/ft^2 per year), or megajoules per square meter per year (MJ/m^2 per year). A note about units of energy: One BTU (British thermal unit) is about the amount of energy expended in burning a standard wooden match.

Designers often look at components of systems for ways to reduce energy consumption. A better approach is to examine the total system. The system starts at incoming power leads, includes the distribution cable, the motor control, the motor, how the motor is connected to the load (such as VFD, flexible coupling, gear train, belt, or direct), the driven load, the control of the load (electronic, pneumatic, or hydraulic), and complex thermodynamic and physical properties heat exchange models (Figure 9.2).

Thus, while building M/E engineers had gained the ability to construct detailed heat and power balances for entire facilities, they had no structured way of evaluating the results for optimality (maximize fuel or power to water/air efficiency). As a result, facilities tended to be broken down into subsystems that could be evaluated easily. Improving processes by optimizing within subsystems, however, frequently results in suboptimal overall design (often not known to the engineer). An integrated approach is really a systematic application of the thermodynamic laws, Ohm's law, Kirchoff's law, Maxwell's law, etc. to building systems.

To calculate power flow to thermal, air, or water flow efficiency requires psychometric, hydraulic, and electrical knowledge. For example, in the case of water, flow-head characteristics and pump performance throughout the total speed range comprise most of the hydraulic while motor and variable speed drive efficiency constitute the electrical information. Thus power to water efficiency is the

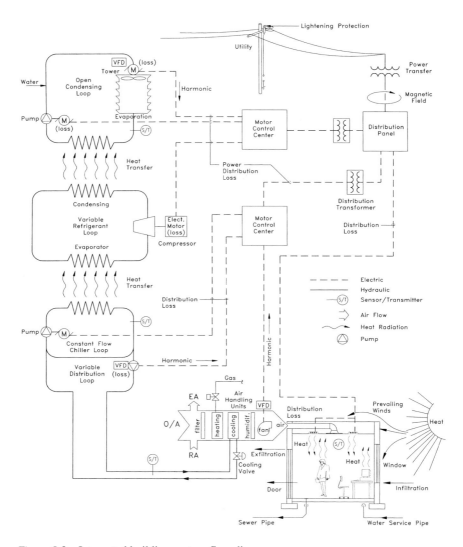

Figure 9.2   Integrated building system flow diagram.

energy transferred to the water by the pump divided by the electrical energy consumed by the motor, the variable speed drive, power filter, wiring, transformer, power distribution equipment, and the utility transformer. The equation for power to water efficiency is:

$$\frac{\text{system flow} \times \text{system head}}{\text{total electrical energy consumed}}$$

This approach can be further carried on to water coil in the air handler to get a better view of the relationships between power, water, and air-side systems. They are not separate. To design them separately is to miss an opportunity to improve system performance. Furthermore, by evaluating each of the elements that contribute to the loss, the individual contributions of power transfer, heat transfer, mass transfer, and momentum (pressure differential) can be illustrated. Electromechanical energy conversion involves energy in at least four forms, and conservation of energy leads to the following relation among these forms:

Energy input from electrical source
 = mechanical energy output
  + increase in energy stored in coupling field
  + energy converted to heat

It is applicable to all conversion devices; it is written so that the electrical and mechanical energy terms have positive values for motor action. This applies equally well to generator action; the electrical and mechanical energy terms then have negative values.

The real challenge, however, is to properly integrate within the design energy, cost and power budget constraints. Having the tools and understanding their application does not ensure success in itself. To achieve real success, the "work process" (functions that must be performed to produce a defined set of deliverables) adopted is as important as the integrations themselves.

9.5.1 Power Budgets

The requirement to establish a project's (building shell) power budget in accordance with a specified procedure has now been incorporated into the building codes of many states. The purpose of this budget determination procedure is not to dictate design procedure. Instead, it is to develop an overall maximum power budget *within* which the designer is free to do as he or she wishes. Obviously, prodigality in one area is necessarily at the expense of another area, since maximum power is inflexible, and the entire budget is built on reasonable design techniques. Still, there is enough leeway in the budget and enough exceptions so that the designer is not overly restricted.

The unit power density (UPD) procedure utilizes precalculated unit power densities in watts per square meter (foot) for specific types of areas and usually standards contain an extensive table of values. Because of the multiplicity of standards and guidelines, the system designer may find a contradiction between interdiscipline authorities. In such an event, written clarification from the appropriate authorities is advisable.

A more complicated set of regulations applies to the design of walls, roofs, and so on—these establish maximum overall thermal transfer values. These more de-

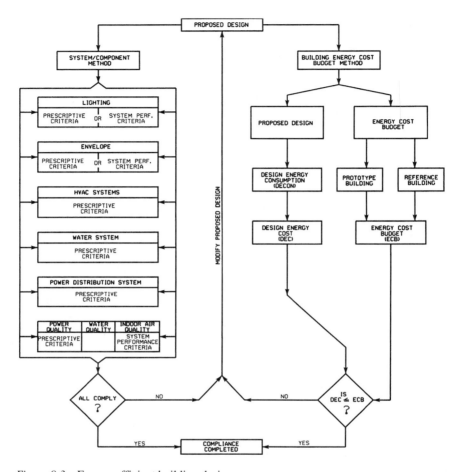

Figure 9.3   Energy-efficient building design.

tailed regulations would also include HVAC system performance (indoor temperatures, controls, ventilation rates, humidity control, zoning, pipe and duct insulation, equipment performance ratings, etc.), domestic hot water efficiency rating, electrical distribution systems, and lighting efficacy. See Figure 9.3 for a design process a system designer needs to use to ensure energy efficient building design.

9.6   Load Analysis

Loads are like king and queen to a building system designer. He or she has to cater to temperamental king and demanding queen. It is extremely important to keep system perspective during the process of calculating loads for buildings. In brief, this stage consists of the following steps:

1. Addressing load (human comfort, process, and equipment) is the ultimate objective of electrical and mechanical calculations and resulting architectural manipulations. Achieving numerical adequacy is important, but it is not by itself a guarantee of accommodating comfort and characteristics of load.

2. Buildings, like our bodies, exchange heat with the outside environment in two distinct ways: *through the envelope,* akin to heat exchange through human skin, and with *incoming fresh air,* akin to our heat exchange with the air we inhale. When calculating envelope (skin) losses or gains, materials, areas, and rates of heat flow through the envelope are used. When calculating fresh air (lung) losses or gains, volumes of space and rates of fresh air exchange are used. For heavily insulated buildings in cold climates, it is not unusual for "lung" losses to exceed "skin" heat losses. Skin and lung losses (or gains) must be added together to determine the total rate of a building's heat exchanges with its environment; both can be manipulated through design to improve a building's thermal performance.

3. Buildings nearly always experience internal heat production; this is helpful in meeting cold weather demands, but harmful in hot weather. For most buildings, it will be helpful to manipulate the envelope to *minimize* internal gains (e.g., to reduce the need for electric lights by specifying larger, better-placed windows), since the reduction in internal heat is so helpful in summer, and can so often be replaced by an increased reliance on passive solar heating in winter.

4. An occupied building can be very different thermally from an unoccupied one. Many buildings (such as offices and schools) experience high internal gains and rates of fresh air for only about a third of a typical day, and only 5 days per week; this amounts to less than a quarter of the year. It is important to remember that although we are designing to provide thermal comfort while a building is being used, we cannot depart too far from comfortable interior temperatures during the night (or weekend) without straining the heating and cooling systems early in the morning. Said another way, one hour's surplus heat can be another hour's needed warmth.

5. Variations in tasks performed inside a building are common in public buildings. For example, a school gym can be used for athletics, band concerts (despite the acoustics), and town meetings—three totally disparate load requirements. In these and similar instances it is common practice to treat the space as essentially three different spaces and to

calculate loads and design systems for each with a careful eye to maximum common equipment usage. Similar problems are encountered in basements, multipurpose rooms, and conference/meeting/lecture/exhibition rooms.

6. Prolonged intensive application or rapidly changing loads would require system design to be raised on (or more) levels. Factors to be considered in addition to the nature of the load are its repetitiveness, variability, condition of the occupants, load duration, and special requirements. All internal loads in a building vary in time and amplitude on basis of usage; thus a diversity factor should be applied to all the loads. Internal load density (ILD) is the sum of the lighting power density (LPD), the equipment power density (EPD), and the occupant load density (OLD):

$$ILD = F_L \, (LPD) + F_E \, (EPD) + F_O \, (OLD)$$

where

F_L = lighting adjustment factor

F_E = equipment adjustment factor

F_O = occupant adjustment factor.

7. Lighting load in most calculations is misrepresented owing to lack of a lighting energy conversion process. If the energy in any light source could be converted without loss into yellow-green light the efficacy of the source would be 683 lumens/watt but three main energy conversions occur during the process of generating light by a fluorescent lamp. During each of these conversions some energy is lost, so that only a small percentage of the input is converted into visible radiation. Ultimately only 21% of input energy is converted to light with 37% to infrared and remaining 42% dissipated heat. (Refer to Figure 3.1.)

These calculations are a rare opportunity to numerically evaluate a building design. When the numbers are all cranked out, sit back and take an *integrated* view. What design strategies do they suggest? We've worked hard toward happiness of king and queen and made them agree with each other; now let them work for us.

9.7 Integrated Design Steps

Building systems usually involve a minimum of three design stages. In the *preliminary design* phase, the most general combinations of comfort and load needs and climate characteristics are considered:

- Activity comfort needs are listed.
- Load schedule is listed.
- Site energy resources are analyzed.
- Climate design strategies are listed.
- Building form alternatives are considered.
- Combinations of passive and active systems are considered.
- One or several alternatives are sized by rule of thumb.

For smaller buildings, this analysis is often done by the architect alone. For innovative or unusual systems in smaller buildings, and especially for larger, multiple-zone buildings, consultants such as engineers and landscape architects often are included. The integrated approach is particularly valuable in assessing the strengths of various design alternatives. The architect and the building system consultants have very different perspectives, and when mutual goals can be clearly agreed upon early in the design process, these perspectives are not only mutually supporting but can produce striking innovations whose benefits extend far beyond services to the clients of a particular building. By setting an example, the team can make available better environments for less energy.

By the time the *design development* phase is reached, one of the design alternatives has probably been chosen as the most promising combination of aesthetics, social, and technical solutions for the program. The consulting engineer (or architect, on a smaller job) is furnished with the latest set of drawings and the program. Typically, the building system engineer then:

1. Establishes design conditions.
   a. By load requirement, lists the range of acceptable air and surface temperatures, air motions, relative humidities, lighting levels, and background noise levels.
   b. Establishes the schedule of operations.
2. Determines the HVAC, lighting, and power zones, considering:
   a. Activities
   b. Schedule
   c. Orientation
   d. Internal heat gains.
3. Estimates the thermal, lighting, and power loads on each zone.
   a. For worst winter conditions
   b. For worst summer conditions
   c. For the average condition or conditions that represent the great majority of the building's operating hours
   d. Frequently, an estimate of annual energy consumption is made.

4. Selects the basic environmental systems. Often, several HVAC systems, lighting configurations will be used within one large building, since orientation, activity, or scheduling differences may dictate different lighting and HVAC solutions. Especially common is one system for the all-interior zones of large buildings and another system for the perimeter zones. Power systems arrangements are more focussed on reliability issues; voltage level is normally dictated by load.

5. Identifies the building system components and their locations:
   a. Mechanical rooms; electrical rooms
   b. Distribution trees (power, air, water, signal)—vertical chases, horizontal runs
   c. In-space components, such as under window fan-coil units, ventilated windows, air grilles, power and lighting panels/controllers, sensors, plumbing fixtures, and so on.

6. Sizes the components.

7. Electrical engineer reviews total load on building and checks against capacity of power system.

8. Although the eventual choice of system should follow an analysis of the zone's needs, some early concepts underlie system choices. The first we will consider is the question of *central versus local* systems. Another basic question is that of *uniformity versus diversity* in the interior environments of buildings. This question encompasses not only thermal experiences, but visual and acoustical ones as well.

9. Lay out the system. At this stage, conflicts with other building components (structure, architectural, plumbing, fire safety, circulation, etc.) are most likely to become evident. Since insufficient vertical clearance is one of the most common building coordination problems with building systems, the layouts must include sections as well as plans. Opportunities for integration with other systems also become more apparent at this stage.

After the architect and the other consultants hold conferences in which building system layout drawings are compared to those for other building components (structural, architectural, plumbing, etc.), *design finalizing* occurs. At this final stage, the system designer verifies the match between the building system equipment operating weight and building structural/architectural component's capacity to meet the load. Final layout drawings then are completed.

The flexibility in building systems arrangements that are accompanied by uniform ceiling heights, light placement, grille locations, and so on, can extend a building's usable life span. However, uniformity is not always attractive to users, and diversity is often encouraged at a more personal level—with office furnishings, for example. A more thorough approach to diversity can provide a stimulus to the user who spends many hours away from the variability of the exterior climate.

Building system lighting designer and architects have long recognized that a space can be made to seem brighter and higher if it is preceded by a dark, low transition space. Thermal comfort impressions can be manipulated similarly. Diversity in the thermal conditions to be maintained, such as less-than-comfortable conditions in circulation spaces or other less critical zones, not only make the critical spaces seem more comfortable by contrast, but also save significant amounts of energy over the life of a building. Further, such conditions can make passive strategies more attractive.

For the benefit of the reader whose background and study framework is purely mechanical or electrical or fire protection or security, it may be helpful to draw parallels between the restrictive approach to building system design.

9.7.1   Evaluation Stage

With the design on paper, it can now be analyzed for conformance to the principal constraints of cost and energy. If the design stage has been carefully accomplished, with due attention to integrating factors, the result of the final evaluation should be gratifying. The results of this stage are fed to the owner group for use in the final overall project evaluation.

Standard, energy-efficient design of new buildings presents two types of design evaluation procedures: the system/component method and the building energy cost budget method. Both methods rely on the use of microcomputers, although there is a quick "prescriptive path" within the system/component that is presented in Figure 9.3.

10

Trends, Recent Shifts, and Some Challenges

Current technology in the computer industry is changing rapidly. By the time a computer system is purchased and implemented, new technology is already on the market that renders it obsolete. This situation, along with the need to share information and applications among various computer systems, makes a system design that can readily adapt to change a necessity. An information system has an architecture in much the same way that a building does. If laid out properly, this architecture can accommodate the introduction of new technologies while allowing the continuing use of existing software.

A computer requires both hardware and software to perform useful tasks. The term hardware refers to the physical equipment (hardware) and the term software refers to programs or sets of instructions (software) that direct the equipment to perform the desired tasks. Software can be divided into four major categories: system software, languages, utilities, and application programs.

System software, otherwise known as the operating system, is the environment in which other programs run. It handles input and output (keyboard, video display, and printer) and file transfer between disks and memory; it also supports the operation of other building system programs. Languages are used to write computer programs. They range from assembly language, which involves coding at the machine instruction level, to high-level languages such as Pascal, C, or C++ which are emerging as preferred standards for professional programming in many industries, including building systems.

The operation and maintenance of buildings has also benefitted from more powerful computers that monitor, control, and in certain cases diagnose problems in building systems equipment.

10.1   Digital Computers as Design Tools

The use of digital computers in the building system industry has come about because of the wide variety of easily used engineering analysis programs for the industry, an even larger number and range of programs for business use, and the low cost of powerful computers on which to run them. The ordinary calculations required in building systems, such as heating and cooling loads, electrical short circuit, and coordination can be performed easily and inexpensively on a computer. In addition, computers sometimes allow the solution of more complicated problems that would otherwise be impractical to solve.

Utility software programs perform standard organizing and data-handling tasks and applications software perform a particular application that require many functions and utilities. For example, *building systems application programs* are designed to use the computer's power to calculate items such as loads, energy, and piping and wiring design.

Many special-purpose application programs exist, including proprietary equipment selection packages and microcomputer versions of public domain programs such as DOE-2, BLAST, TRNSYS, EDSA, DAPPER, and other powerful simulation programs previously available only on large mainframe computers. Because computers do repetitive calculations rapidly, accurately, and tirelessly, it is possible for the designer to explore a wider range of alternatives and to use selection criteria based on annual energy costs or life-cycle costs, which would be much too tedious without a computer.

Characteristics of a Loads Program: In general, a building HVAC system loads program requires user input for most or all of the following:

- Full building description, including the construction of the walls, roof, windows, etc., and the geometry of the rooms, zones, and building. Shading geometries may also be included
- Sensible and latent internal loads due to lights and equipment, and their corresponding operating schedules
- Sensible and latent internal loads due to people
- Indoor and outdoor design conditions
- Geographic data such as latitude and elevation
- Ventilation requirements and amount of infiltration.

With this input, loads programs will calculate both the heating and cooling loads as well as perform a psychometric analysis. Output typically includes peak room and zone loads, supply air quantities, and total system (coil) loads.

The calculation of heating and cooling loads is dependent on the choice of weather data. Heat extraction is the rate at which the building system removes heat from a conditioned space. When the space temperature is kept constant, this rate equals the cooling load; because this rarely happens, heat extraction is generally either smaller or larger than the cooling load. This concept is important for the analysis of intermittently operated building systems; it provides information on the relationship between load and space temperature and leads to the calculation of the preheat/cool load for various combinations of equipment capacity and preheat/cool periods.

10.2   Computers and Integrated System Simulation

Although computers are now widely used in the design process, most programs perform not design, but simulation. That is, the engineer proposes a design and the computer program calculates the consequences of that design. When alternatives are easily cataloged, a program may design by simulating a range of alternatives and then selecting the best according to predetermined criteria. Thus, a program to calculate the annual energy usage of a building requires specifications for the building and its systems; it then simulates the performance of that building under certain conditions of weather, occupancy, and scheduling. A building system program may actually size duct, pipe or wire, but usually an engineer must still decide quantities, routing, and so forth.

10.2.1   Building Energy Simulation

Building energy simulation programs differ from peak load calculation programs; in the former, loads are integrated over time (usually a year), the systems serving the loads are considered, and the energy required by the equipment to support the system is usually the calculated output. Most energy programs simulate the performance of already designed systems, although programs are now available that make selections formerly left to the designer, such as equipment sizes, system air volume, and fan power. Energy programs are fundamental in making decisions regarding building energy use and, along with life-cycle costing routines, quantify the impact of proposed energy conservation measures during the design phase. In new building design, energy programs help determine the appropriate type and size of building systems and components; they also explore the effects of design tradeoffs and can be used to evaluate the benefits of innovative control strategies and the efficiency of new equipment.

Energy programs that track building energy use accurately can help in determining whether a building is operating efficiently or wastefully. They have also been used to allocate costs from a central heating/cooling plant among buildings served by the plant. However, one must be certain that such programs have been adequately calibrated to measured data from the building under consideration.

Characteristics of Building Energy Simulation Programs: Most programs simulate a wide range of buildings, mechanical equipment, and control options. While the importance of approximating solar loads on sunny days is widely recognized, computational results differ substantially from program to program. The shading effect from overhangs, side projections, and adjacent buildings is frequently a factor in the energy consumption of a building; however, the diversity of approaches to the load calculation results in a wide range of answers. Depending on the requirements of each program, various weather data are used. These can be broken down into five groups:

1. Typical hourly data for one year only, from averaged weather data
2. Typical occupancy data for one year, as well as design conditions for typical design days
3. Reduced data, commonly a typical day or days per month for the year
4. Typical reduced data, nonserial or bin format
5. Actual hourly data, recorded on-site or nearby, for analysis where the simulation is being compared to actual utility billing data or measured hourly data.

Both air side and water side energy conversion simulations are required to handle the wide variations among central heating, ventilation, and air conditioning systems. For proper estimation of energy use, simulations must be performed for each combination of system design, operating scheme, and control sequence.

10.2.2 System Simulation Techniques

Two basic approaches are currently used in computer simulation of energy systems: the fixed schematic technique and the component relation technique. The fixed-schematic-with-options technique is the most prevalent program organization. Used in the development of the first generally available energy analysis programs, this technique involves writing a calculation procedure that defines a given set of systems. The system schematic is then fixed, with the user's options usually limited to equipment performance characteristics, fuel types, and the choice of certain components.

Component Relation Technique: Advances in system simulation and increased interest in special and innovative systems have prompted interest in the component approach to system simulation. The component relation technique differs from the fixed schematic in that it is organized around components rather than systems. Each of the components is described mathematically and placed in a library for use in constructing the system. The user input includes the definition of the system schematic, as well as equipment characteristics and capacities. Once all the components of a system have been identified and a mathematical model for each has been formulated, the components may be connected together in the desired manner, and information may be transferred between them. Although certain inefficiencies are built into this approach because of its more general organization, the component relation technique does offer versatility in defining system configurations.

Data on building occupancy are among the most difficult to obtain. Since most energy analysis computer programs simulate the building on an hourly basis for a one-year period, it is necessary to know how the building is used for each of those hours. Frequent observation of the building during days, nights, and weekends shows which energy-consuming systems are being used and to what degree. Measured, submetered hourly data for at least one week are necessary for a beginning of an understanding of weekday/weekend schedule-dependent loads.

10.2.3   Equipment Selection and Simulation

Three types of equipment-related computer programs are equipment selection, equipment optimization, and equipment simulation programs. Equipment selection programs are basically computerized catalogs. The program locates an existing equipment model that satisfies the entered criteria. The output is a model number, performance data, and sometimes alternative selections.

Equipment optimization programs display all possible equipment alternatives and let the user establish ranges of performance data or first cost to narrow the selection. The user continues to narrow the performance ranges until the best selection is found. The performance data used for optimizing selections vary by product family.

Equipment simulation programs calculate the full- and part-load performance of specific equipment over time, generally one year. The calculated performance is matched against an equipment load profile to determine energy requirements per hour. Utility rate structures and related economic data are then used to project equipment operating cost, life-cycle cost, and comparative payback.

Some advantages of equipment programs include the following:

- High speed and accuracy of the selection procedure
- Pertinent data presented in an orderly fashion
- More consistent selections than with manual procedures
- More extensive selection capability
- Multiple or alternate solutions
- Small changes in specifications or operating parameters easily and quickly evaluated
- Data-sharing with other programs, such as spreadsheets and CAD.

Simulation programs have the advantage of (1) projecting part-load performance quickly and accurately, (2) establishing minimum part-load performance, and (3) projecting operating costs and payback-associated higher performance product options.

There are programs for nearly every type of building system equipment, and industry standards apply to the selection of many types. The more common programs and their optimization parameters include the following:

Air distribution units	Pressure drop, first cost, sound, throw
Air handling units Rooftop units	Power, first cost, sound, filtration, heating, and cooling capacity
Boilers	First cost, efficiency, stack loses
Cooling towers	First cost, design capacity, power, flow rate, air wet bulb temperatures
Chillers	Condenser head, evaporator head, capacity, power input, first cost, compressor size, evaporator size, condenser size, refrigerant type
Coils	Capacity, first cost, water pressure drop, air pressure drop, rows, fin spacing
Fan coils	Capacity, first cost, sound, power
Fans	Volume flow, power, sound, first cost, minimum volume flow
Heat recovery equipment	Capacity, first cost, air pressure drop, water pressure drop (if used), effectiveness
Pumps	Capacity, heat, impeller size, first cost, power
Air terminal units (variable and constant volume flow)	Volume flow rate, air pressure drop, sound, first cost
Lighting	Illuminance, efficacy, fixture type, spacing and quantity selection based on foot candles required.
Power Systems	Load flow, voltage drop, short circuit and protective devices coordination, power factor correction, resistance/reactance/impedance network, power demand
Switch boards, panel boards	Voltage, phase, current interrupting rating, surge protection, insulation and protection options

| Transformers | Primary and secondary voltage, phase, winding configuration, impedance, efficiency, insulation rating |

Chiller, refrigeration equipment, and transformer and motor selection programs can choose optimal equipment based on such factors as lowest first cost, highest efficiency, best load factor, and best life-cycle performance. In addition, some manufacturers offer modular equipment for customization of their product. This type of equipment is ideal for computer selection. Chiller selection programs now include the type of refrigerant and its ozone depletion potential.

However, equipment selection programs have limitations. The logic of most manufacturers' programs is proprietary and not available to the user. All programs incorporate built-in approximations or assumptions, some of which may not be known to the user. Equipment selection programs shall be qualified before use.

10.3   Computer Graphics and Modeling

Graphics and Imaging Software: Microcomputer software graphics packages readily produce x–y and bar graphs, pie charts, and CAD drawings, and often have a library of graphic elements that make possible the rapid creation of custom presentations. The new technology of multimedia tools allows video and sound to be combined with traditional text and graphics.

Combining alphanumerics with computer graphics enables the design engineer to quickly evaluate design alternatives. Computers can pictorially simulate design problems with three-dimensional color displays that can predict the performance of environmental systems before they are constructed. Simulation graphics software is also used to evaluate design conditions that cannot ordinarily be tested with scale models owing to high costs or time constraints. Integrating computer graphics software with expert system software can further enhance design capability. Integrated CAD and expert systems can help the building designer and planner with construction design drawings and with construction simulation for planning and scheduling complex building construction scenarios.

10.4   Communication and Artificial Intelligence

Communication between computers takes place through telephone lines or other networks of various configurations. Coaxial cables, twisted pair wiring, superimposed ac carrier frequencies, fiber optics, radio transmissions, and infrared light are among the media used. Features available on such networks include bulletin boards, on-line retrieval services, shared services, electronic mail, and file transfer.

Electronic communication using existing telephone lines is rapidly becoming possible anywhere there is telephone service, even overseas exchanges or with cellular phones. Telephone systems around the world are being upgraded to transfer digital signals, which will vastly expand capabilities. Electronic communication continues to require both appropriate software and a modem, which performs the translation of the digital computer signals into analog signals suitable for current telephone lines. The modems most commonly used at present transfer data at 1200, 2400, 4800, 9600, or 14,400 bits per second (bos) using one of several transmission standards.

10.4.1 Applications of Artificial Intelligence

There are many different applications of artificial intelligence (AI), ranging from systems that learn, reason, and recognize speech like human beings to those that can help make decisions based on incomplete, uncertain, and/or complicated information. In building systems artificial intelligence techniques are generally used in three areas—design, control, and diagnosis.

Knowledge-Based Systems: Knowledge-based systems (KBSs), often called expert systems, can make decision similar to those made by human experts. KBSs typically use facts, if–then rules, and/or models to make decisions. Most KBS tools can also link with other programs, access databases, and import graphics. Expert systems shells are usually complete development packages that include rule and database development tools, debugging facilities, and good user interfaces. Some shells take advantage of windowing environments and incorporate object-oriented features, hypertext, and graphics.

Artificial Neural Networks: Artificial neural networks (ANNs) are collections of small individual interconnected processing units. Information is passed between these units along interconnections. An incoming connection has two values associated with it: an input value and a weight. The output of the unit is a function of the summed value. ANNs, while implemented on computers, are not programmed to perform specific tasks. Instead, they are trained with repeated data sets until they learn the patterns presented to them. Once the net is trained, new patterns may be presented to it for classification.

Fuzzy Logic: Fuzzy logic is a form of mathematics that allows precise computations using inexact and uncertain input. It operates by assigning a membership value in a fuzzy set to the values of each input. For example, the set of (membership value, fuzzy value) for a temperature of 71.6 °F might be [(0.0, frigid), (0.1, cold), (0.7, comfortable), (0.3, warm), (0.0, hot)]. The membership values do not have to add up to 1.0. Mathematics and rules are then applied to the fuzzy values (frigid, cold, comfortable, warm, hot) to arrive at the conclusion, which is the output fuzzy set. The output fuzzy set is defuzzyfied, using a centroid method, to produce a real value if needed. Building systems applications of fuzzy logic include chiller controls and motor control systems.

Genetic Algorithms: Genetic algorithms (GAs) are search methods based on the mechanics of natural selection. Each element of a string represents a parameter in the process being optimized. Each set of strings represents a possible solution for the process. Genetic algorithms differ from traditional search and optimization techniques in that they manipulate the coding of parameter sets, not the parameters themselves. They search for the optimal solution for a number of points, rather than for a single point, using probabilistic rather than deterministic rules. There are a limited number of applications today that use GAs; most of these are in the aerospace industry.

11

Acoustics, Vibrations, and Integrated Building Systems

11.1 Introduction

Sound is transmitted by waves. These waves travel through gases, liquids, and solids alike. Wave action is mainly a transfer of energy. In the case of sound, this transfer takes the form of tiny motions at the molecular level. In gases and liquids, a sound wave shifts molecules slightly in a direction parallel to itself, that is, in a lengthwise direction. In solids, motion may also occur perpendicular to the wave. Sound waves spread from their source in all directions. The speed of sound is independent of loudness of tone. The sounds from a radio in a room, whether they are loud or soft, of high pitch or low, all reach a listener simultaneously.

Mechanical oscillation occurs in a broad range of frequencies. Since an adult can hear oscillations between approximately 100 cycles per second and 15,000 cycles per second, we call that range audible sound. Below the audible sound range, we call oscillation vibration, and above the sound range we call oscillation ultrasound or ultrasonics. Vibration, audible sound, and ultrasound are exactly the same thing except for frequency.

Ultrasound is conducted by the air, although humans do not hear it. The burners of a hot air balloon generate enough ultrasound to drive dogs on the ground into a frenzy. And, of course, there are ultrasonic dog whistles. Ultrasound is severely damped by most materials, so there is not much of a problem in providing barriers. In fact, the principal mechanical design problems involving ultrasound are to provide conduits for it.

The ideal building acoustical environment would permit any occupant to talk easily or talk on a telephone without distracting or being understood by other occupants. The best way to design such an environment is to look at building systems as a man–machine system. This system may be considered in three principle parts:

Noise Sources: Building systems electrical and mechanical equipment; flow of liquid in pipes, air in ducts or pipes, current in conductors, or flux in magnetics; people who talk and move about; office equipment; and anything else that is likely to make annoying noises.

Noise Transmission Elements: Constitute the bulk of the architectural elements of building system, such as ceilings, walls, floors, screens, furniture layout, and anything that influences the amount of noise reaching the receiver.

Noise Receiver: Predominantly people with a common, but not completely singular, sensitivity to these noises. Their sensitivity may be modified by adding masking sound, which does not annoy them, to their environment.

11.2   Acoustical Environment of Buildings

It is the responsibility of the system designer to provide the proper environment by confining noise sources under his or her control, achieving the best attenuation along noise transmission paths, and understanding the sensitivity of the people-receivers. The bulk of quantitative knowledge developed to date centers on the second item—attenuating, absorbing, canceling, and masking subsystems. Not much has been quantified on receiver sensitivity in this context.

Interfering Sounds: Common office equipment beside building systems can be equally annoying. Sometimes the most irritating noise in an open area is the clamoring telephone bell. Other machines, such as copiers, tend to be less numerous and may be given special placement to minimize their noise. Other sources of mechanical noise, such as ventilation systems, should be kept below noise criteria curve 35 (NC35) (or approximately 45 dBA) throughout the office. This is not a stringent requirement and is usually achievable. Noise control procedures are generally most effective when applied directly to the source.

Building System Noise: Mechanical devices obviously make noise, and, generally, the more power they consume, the more noise they make. In many of today's buildings, 40% of the total cost is spent on mechanical systems. These systems are located throughout a building.

In most buildings, the primary sources of mechanical noise are the components of the air conditioning and air handling systems such as fans, compressors, cooling towers, condensers, ductwork, dampers, mixing boxes, induction units, and diffusers. Figure 11.1 shows noise and vibration sources of a rooftop unit commonly used in commercial buildings. The increased use of variable air volume (VAV) systems has introduced some noise problems that should not be neglected. VAV system noise can be minimized by maintaining minimum system static pressure since fan noise increases exponentially with static pressure. Ceiling diffuser acoustic characteristics must be integrated to

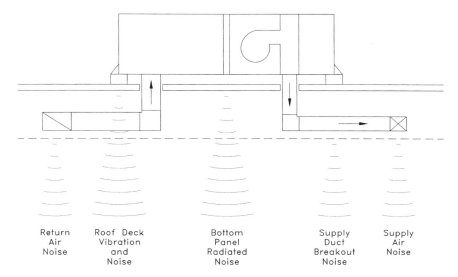

Figure 11.1   Rooftop unit noise sources.

meet requirement of design flow and masking noise. Normally piping is not a major noise source, but where flow velocities are high and the pipe is in contact with the structure, it can create noise problems. Pumps are another source of mechanical noise. Pump noise is frequently transmitted along pipes to remote points.

11.3   Noise and Acoustic Planning Issues

Too often, building systems are designed with little concern for controlling noise and vibration. Design team members may ask equipment vendors for their opinions during the design process, or an acoustical consultant may be retained just before the drawings are released for bids to check things over and make last-minute recommendations. In such instances, the site plan, space plan, equipment selections, duct and pipe routing, and structural design are all complete and are unlikely to be changed.

   On the other hand, if the project's acoustical aspects are given adequate priority during the early design phases, the design team has the necessary flexibility to select and locate HVAC equipment for proper operation and effective noise and vibration control. In the early design phases, this is done by effective site planning, space planning, equipment selection, and equipment room sizing. Proper consideration of acoustics early in the design may allow solutions to potential problems at little or no cost.

This planning issue is important in the placement of ground-based, outdoor HVAC equipment such as cooling towers, air-cooled condensers, and air-cooled chillers. Many cities and counties have ordinances that place limits on the permitted level of building support system noise at adjacent property lines. Complying involves keeping noisy equipment away from the property line, thereby bringing the equipment closer to the building that it serves. This planning option must consider whether or not the equipment's proximity to the building may cause interior noise problems from sound penetration through the building's windows, doors, or ventilation openings.

The mechanical engineer should make tentative equipment selections as early as possible to allow for a preliminary noise analysis and to determine the probable sizes of the mechanical rooms. The primary acoustical design guideline is to select the quietest equipment possible. This usually means selecting equipment that will perform at its most efficient operating point. For fan-based equipment, select a fan with the lowest practical rpm, while avoiding proximity to the fan's stall region for all expected operating conditions.

After the equipment locations have been determined and the tentative selections have been made, the equipment locations and operating weights should be given to the structural engineer. The structural engineer will use the information to help select beam sizes and column spacing, aspects of the design that are very difficult to change later.

Economic pressure to maximize a building's rentable floor area has resulted in less space being available for the HVAC system and other building services. This trend often forces the mechanical engineer to specify small, inefficient equipment or to shoehorn properly sized equipment into a restricted space. Both options can lead to excessive noise in neighboring spaces, which may then become nonrentable.

Electrical equipment is generally overlooked as a noise source and this is unwise and a drawback of not integrating acoustic design concerns. Most electrical equipment noise is a 120-Hz hum on a 60 Hz system or 100 Hz on a 50 Hz system. This can be very disturbing because it is so low a frequency, and low-frequency noise is difficult to attenuate.

A typical motor/blower exhibits mechanical, aerodynamic, and electromagnetic noise. Its level and frequency are directly proportional to motor speed. The noise spectrum typically will include both discrete frequency components (tones) and random broad-band sounds. The bearings are the primary sources of mechanical noise in electric motors together with any unbalances of the rotor. A well-balanced rotor operating below its critical bending speed (operating frequency coincident with the bending frequency of the shaft) is essential for quiet high-speed motor operation.

However, all rotating elements have some degree of unbalance, characterized by a sinusoidal vibration that may be transformed into acoustical energy, possibly at a frequency of once per revolution. The aerodynamic noise of concern is produced by the impeller (rotor), which generates a frequency spectrum containing the two major components, discrete tonal noise and its broad-band counterpart:

- Tonal noise is produced when the air leaving the equally spaced blades hits a stationary obstacle near the impeller. The tonal frequency is commonly called the blade passing frequency (BPF).
- Broad-band noise is generated by turbulence and other fluid instabilities of the air flow over the blades, plus interaction with local structures. It is characterized by equal amounts of energy across the audible frequency range which usually dominates the radiated sound power spectrum of the motor/blower.

In addition, the electrical noise such as that created as the rotor slot passes a stator air gap is produced by the electric motor. However, these levels are small in comparison to those created by the motor/blower. Another major source of 120-Hz hum is conventional, core-and-coil discharge light fixture ballasts. (Electronic ballasts are practically noiseless.) This includes fluorescent plus all the HID sources.

11.4 Masking Sound

This is commonly used or applied in open office areas where personnel are separated by partitions. An individual speaker in an average office or work area will normally modulate his voice so that his listener will enjoy a nearly 20 dB signal-to-noise ratio at his listening distance, typically 1.2 m (4 ft).

In addition to noise transmission through or diffraction about building partitions, intruding sounds may reach the listener by bouncing from nearby flat surfaces. This process is termed "specular reflection" and is analogous to the bounce of a billiard ball from a cushion or the reflection of light by a mirror. The amplitude of this reflected sound depends on the reflective properties of the surface. Reflectivity is not always predictable by the noise reduction coefficient (NRC) of the surface, primarily because NRC averages absorptivity at all angles of incident sound.

Privacy Criteria: At the onset of the design process, it is important to determine the need for sound level, since measures necessary to provide acoustic privacy later can be expensive. The noise criterion for a given function depends on that function. Executive and sensitive areas require confidential privacy, defined as zero phase intelligibility. Other areas may require minimal privacy, such as secretarial pools. To date it has been customary to relate privacy requirements to speech communication as discussed in the following sections.

Speech Sounds in the Open Office: Average male conversational speech at a distance of 1 m (3 ft) has a broad-band level of 60 to 65 dB. The dynamic range of component levels is 30 dB across the frequency range from 200 to 5000 Hz. These limits are not firm. It is usually accepted that the middle frequency components are more important for communication than the low and high frequencies. For 100% intelligibility, the full 30 dB dynamic range of the complex frequency spectrum should lie above the prevailing background noise. Intelligibility of speech is a function of the signal-to-noise ratio. Apart from the dynamic range of 30 dB, the mean value of a normal voice sound has a range of 20 dB from "quiet" through "normal" to "raised," depending on the amount of speech effort applied. Merely raising one's voice may destroy the state of acoustical privacy at nearby work stations. This is perhaps the most irritating human disturbance in the open-plan office. Fortunately the higher absorptive surroundings in the open-plan office appear to induce lower voice levels.

It is common, therefore, to use electronic random noise generators feeding loudspeakers hidden in the ceiling plenum to raise the level of the background. The spectrum of this sound is shaped to effectively mask speech with the least distraction. Figure 11.2 is a block diagram of an electronic masking sound and paging system. Signals from the sound generator are shaped with narrow band filters (for example, one-third or one-half octave) into the spectrum (as heard by the listener) that is considered optimum.

Continuous background noises provided by air conditioning systems, though an attractive expedient, have not been generally successful because of the difficulty in generating the right level with the preferred spectrum. Electronic generators are more flexible. Likewise, music alone is unacceptable because it has pure tone components and transient level characteristics; however, music has been successfully introduced as a supplement to a properly adjusted masking system.

The masking sound field produced in the office must be uniform in space and time to avoid occupant complaints or detection of the masking system. The uniformity of the sound field is governed by the spacing and orientation of loudspeakers, transmission characteristics of ceiling systems, reflective properties of the plenum surfaces, and openings in the ceiling such as plenum air return slots and light fixtures.

A common arrangement is to distribute individual loudspeakers in a square or triangular array with approximately 3 to 6 m (10 to 20 ft) spacing between units. The exact distribution will be influenced by plenum depth, obstructions, ceiling tile properties, and speaker radiation patterns. Ceiling materials that are relatively transparent to sound sometimes use an impervious layer on the back side to diffuse the sound and produce a uniform sound field in the listener areas. Another approach is to mount the speakers over fairly large baffles, about 1.25 m (4 ft) square, so that the sound must reflect and diffract around them, producing a more uniform field.

The monitoring and adjustments should be carried out in the fully furnished office area minus staff. The level chosen for this background sound will depend on the degree of privacy required and on the physical properties of the screens

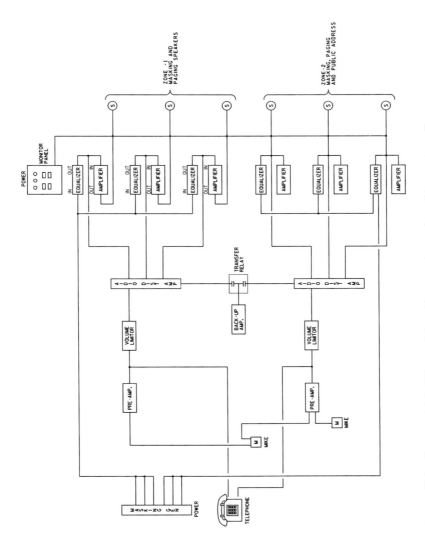

Figure 11.2 Commercial building public address paging and masking system. Two zones are shown for clarity.

and ceiling, although there is an upper limit. Clearly, the introduction of masking sound must not arouse hostility or adverse reactions in the office. Although various occupants have different reactions, most people seem to accept broad-band, steady-state sounds having A-weighted levels up to 52 dBA.

It is best not to have other areas in the building quieter than the open-plan office, lest occupants become aware that their work area is not as quiet as it might be. This suggests the use of masking sound throughout the building. Some work areas require more masking sound level than others, however, and a compromise must be found. One successful technique is a gradual transition zone between levels. A side benefit of the electronic masking system is that it amplifies and loudspeakers can be used for public address announcements, life safety programs, or emergency paging.

11.5   Acoustic Analysis

All sound generation mechanisms and sound transmission paths are potential candidates for analysis. Adding to the computational work load is the necessity of extending the analysis over, at a minimum, octave bands 1 through 8 (63 Hz through 8 kHz). A computer can save a great amount of time and difficulty in the analysis of any noise situation, but the system designer should be wary of using unfamiliar software.

Caution and critical acceptance of analytical results are mandatory at all frequencies, but particularly at low frequencies. Not all manufacturers of equipment and sound control devices provide data below 125 Hz. Thus, the system designer conducting the analysis and the programmer developing the software must make assumptions based on experience for these critical low-frequency ranges.

The designer/analyst should be well satisfied if predictions are within 5 dB of field-measured results. In the low-frequency "rumble" regions, results within 10 dB are often as accurate as can be expected, particularly in areas of fan discharge. Conservative analysis and application of the results is necessary, especially if the acoustic environment of the space being served is critical.

Noise computations for systems should account not only for obvious noise generators such as fans and transformers, but also for the potential for flow generation in duct elements (e.g., trees and branches) and downstream elements (e.g., VAV boxes). Whenever flow velocities exceed 1500 fpm, flow-generated noise is a significant possibility. Sound in ducts propagates upstream as well as downstream, and sound from the fan of a single system travels down at least four paths.

The sound pressure level in a space is the composite of all sound paths to the space. For example, one diffuser in a room may meet noise criteria curve NC-30, but the combined sound pressure level of ten diffusers along with transformer hum in a room may be NC-40 or greater.

Several currently available acoustics programs are generally easy to use but are often less detailed than the custom programs developed by acoustic consultants for their own use. Acoustics programs are designed for comparative sound studies and allow the design of a comparatively quiet system. Acoustic analysis should address the following key areas of the building system:

- Sound generation by HVAC equipment
- Sound generated by passive equipment
- Sound attenuation and regeneration in duct elements
- Wall and floor sound attenuation
- Ceiling sound attenuation
- Sound break-out or break-in ducts or casings
- Room absorption effect (relation of sound power criteria to sound pressure experienced).

Algorithm-based programs are preferred because they cover more situations. However, assumptions are an essential ingredient of algorithms. These basic algorithms, along with sound data from the acoustics laboratories of equipment manufacturers, are incorporated to various degrees in acoustics programs. Whenever possible, equipment sound power data by octave band (including 32 Hz and 63 Hz) should be obtained for the path under study. A good sound prediction program relates all performance data.

Many other more specialized acoustics programs are available. Various manufacturers provide equipment selection programs that not only select the optimum equipment for a specific application, but also provide associated sound power data by octave bands. These programs can help in the design of a specific aspect of a job. Data from these programs should be incorporated in the general acoustic analysis. For example, duct design programs may contain sound predictions for discharge airborne sound based on the discharge sound power of the fans, noise generation/attenuation of duct fittings, attenuation and end reflections of variable air volume (VAV) terminals, attenuation of ceiling tile, and room effect. VAV terminal selection programs generally contain subprograms that estimate the space NC level near the VAV unit in the occupied space. However, projected space NC levels alone are not acceptable substitutes for octave-band data. The designer/analyst should be aware of assumptions, such as room effect, made by the manufacturer in the presentation of acoustical data.

Predictive acoustic software allows system designers to look at HVAC-generated sound in a realistic, affordable time frame. HVAC-oriented acoustic consultants generally assist designers by providing cost-effective sound control ideas for sound-critical applications. A well-executed analysis of the various components and sound paths enables the designer to assess the relative importance of each and to direct corrective measures, where necessary, to the most critical areas. Computer-generated results should supplement the designer's skills, not re-place them.

11.6   Solutions

In a normal reverberant field room where sound undergoes many reflections, the sound field caused by any single source reaches a more or less uniform level at some distance from the source. This remote sound level does not then change ap-preciably from one part of the room to another. On the other hand, sounds cre-ated out-of-doors or in a large open space decrease in level by 6 dB each time the distance from the source is doubled. A well-designed acoustic plan approaches this condition. This is achieved by treating sound reflecting surfaces, especially the ceiling and walls, with absorbing materials to reduce sound reflections to negligible values in the speech frequency range.

Ceiling: The largest surface available for specular reflection of sound in the open spaces is the ceiling. For that reason, much attention is paid to its acoustical properties. The ceiling surface must be highly sound absorbent, especially at an-gles of incidence of 45 to 60°, for a flat ceiling, lest reflections greatly reduce partition effectiveness. The acoustic system designer should provide an ab-sorbent ceiling that approaches an "open sky" condition as nearly as possible.

A variety of ceiling products and configurations are available to reduce ceiling reflections. They range from flat absorbing panels to coffered systems and baf-fles. Accessory installations in the ceiling also can cause sound reflections. Flat, hard surfaces, such as light fixture shielding media, are generally unsatisfactory when located uniformly throughout the ceiling area. An exception to this rule can occur with task-oriented or daylighting.

Walls: Wall carpeting and drapes, although aesthetically pleasing and durable, may not be sufficiently sound absorbing unless extremely thick. Also, they may not have proper fire resistance. It is suggested that acoustically absorptive wall panels specifically designed for acoustic application be utilized. Windows around the perimeter of the building present a problem since most acoustical ab-sorbers will impair vision. Draperies do not provide enough sound absorption unless they are of heavy textile material, fuller than usual, and closed. One possi-ble solution is to tilt the window glass out of the vertical plane, thus deflecting the unwanted sound reflections to the ceiling or floor. Acceptable results may also be obtained by using vertical louvers that do not seriously impede vision but adequately diffuse the otherwise specular reflections.

Floors: The acoustical efficiency of carpeted floors for reduction of impact sounds is well known and highly desirable. The sound absorption of carpeting is of limited value for open-space applications. Since sounds reflected from car-peted floors are easily absorbed by the ceiling system and otherwise broadly dif-fused by the partitions and furniture, no great problems result.

11.6.1 Static Noise Control

The most effective noise control step is to locate noisy equipment and its related services as far as possible from noise-sensitive areas. Nonsensitive areas (such as corridors, storage rooms, toilets, shafts, etc.) can be used effectively as buffer zones to shield noise-sensitive areas from mechanical equipment rooms in a typical multistory office building core area. Optimal mechanical room placement relative to other core areas permits the routing of the return air path and the supply air trunk ducts over areas with low acoustical sensitivity (such as restrooms).

To minimize the chances of noise problems, fan and air handling unit rooms should have floor areas of at least 10 to 15 ft² for each 1000 cfm of equipment air flow. This usually allows adequate space for proper air flow into the fan, low noise supply, and return air duct fittings and duct silencers, if required (Figure 11.3). To minimize further the acoustical coupling between equipment housings and equipment room walls, all building system equipment room should have a floor area large enough to allow a clearance of at least 2 ft (600 mm) around all equipment. In some cases, electrical codes mandate even larger clearances.

Electrical transformer noise can be minimized by these steps:

1. Mount unit on vibration isolators.
2. If transformer is wall hung, use resilient hanging. If it is floor mounted, place on as massive a slab as possible.

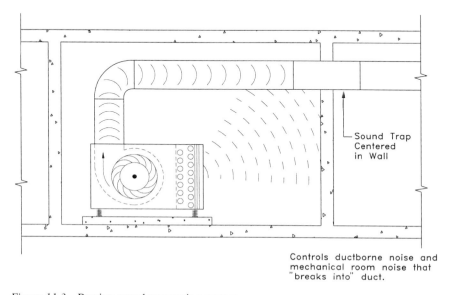

Sound Trap
Centered
in Wall

Controls ductborne noise and
mechanical room noise that
"breaks into" duct.

Figure 11.3   Passive sound attenuation system.

3. Locate the unit so that reflections do not amplify the sound. Sound-absorbent material on the walls behind the units is not useful at 120 Hz. Only cavity resonators will absorb appreciable amounts of sound at that frequency.

4. Use only flexible conduit connections.

5. Avoid locating transformers adjacent to, or immediately outside, quiet areas. A common error in this regard is placing a transformer pad immediately below the window of an NC 15–25 area.

11.6.2   Active Noise Control

Active noise solutions (ANS) uses artificially produced "noise" to counteract unwanted noise. It works by mixing antiphase sounds with the existing in-phase noise to produce "destructive interference" throughout the sound field. The level of attenuation achieved depends on whether the amplitude and phase error of the noise and antinoise signals are matched. When carefully controlled, noise reduction can be as high as 20 dB down.

There are two main approaches to the application of active noise cancellation (ANC). The first is to synthesize a secondary waveform with or without prior information on the original noise signal, which is, however, ineffective against random broad-band noise. To reduce both tones and broad-band noise, a second approach senses the original noise and produces an antinoise signal by filtering and injecting it back into the noise field.

The ANC system consists of three basic components: a controller to produce the antinoise signal, two sensors to measure the combined noise and antinoise sound field, and an actuator to introduce the antinoise into the sound field. Figure 11.4 shows a typical layout of ANC components in a duct. The first sensor microphone is located at a position close to the primary source of the noise, the motor/blower. A second microphone, called the residual sensor, is near the discharge of the duct and an actuator, actually a small loudspeaker, is mounted on the duct wall near the residual sensor. The sensors provide input signal to the controller and the actuator receives controller output. The system's controller is based on a digital signal processor (DSP) chip with appropriate analog input/output circuits for signal conditioning. Software that carries the adaptive feed forward (AFF) algorithm allows this unit to attenuate both broad-band random noise and "periodic" or tonal noise for a variety of applications. The ANC process can be described by explaining how much each component functions during the cancellation cycle. During each DSP clock period, the controller samples both microphone inputs and provides a speaker output. The primary noise sensor sample is filtered and delayed so that when the noise propagates down the duct to the vicinity of the loudspeaker, the signal outputs to the loudspeaker at 180° out of phase. The residual microphone sample provides the means for the

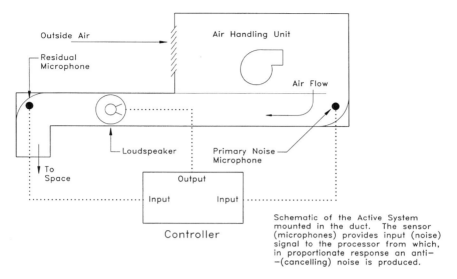

Outside Air

Air Handling Unit

Residual
Microphone

Air Flow

Loudspeaker

Primary Noise
Microphone

To
Space

Output

Input   Input

Schematic of the Active System
mounted in the duct. The sensor
(microphones) provides input (noise)
signal to the processor from which,
in proportionate response an anti-
-(cancelling) noise is produced.

Controller

Figure 11.4   Active sound cancellation system.

controller to monitor and improve its performance by sensing and minimizing the combined sound field of noise and antinoise. The AFF algorithm adaptively computes the cancellation signal and the antinoise signal continuously changes (active) as the noise changes. Errors (residual noise after the introduction of the antinoise signal) are corrected on the following noise cycle.

Several criteria must be satisfied for the ANC system to perform well. The sensors must be at an adequate distance from the actuator to provide a time delay for filtering of the primary noise signal and for acoustic mixing of the combined noise and antinoise field. Second, as the controller is minimizing the error at the residual sensor (using "knowledge" of the primary noise and the cancellation signal), it is necessary that the former's signal characteristics do not change as the noise travels down the duct. Thus, without the loudspeaker signal, both sensors should hear the same noise. This property is called coherence, which is not affected by the time delay between the positions of the primary and residual sensors. However, coherence is greatly affected by turbulent airflow near the microphones. Turbulent aerodynamic noise degrades coherence because it is localized and differs at each microphone. The coherence between microphones sets a theoretical limit on the maximum amount of attenuation that an active cancellation system can achieve. For the motor/blower test platform, this means that duct design, microphone placement, and microphone cover are aimed at preserving coherence across the frequencies where active attenuation would be most effective. (Perfect coherence has a value of 1.0.)

A third criterion for effective noise attenuation is that the controller must have accurate models of the acoustic propagation paths between the actuator and both sensors. These models are obtained during self-calibration wherein the AFF algorithm computes the transfer functions (the time, phase, and amplitude relationship between the output and input of each sensor and the actuator). In the process, two of the digital filters that the AFF algorithm uses are defined. The filter that models the feed-forward electrical and acoustical paths from controller output, through the actuator, to the residual sensor, and back to the controller input is vital to the monitoring and improvement of the adaptive nature of the AFF algorithm. A second filter models the feedback path from the actuator to the primary noise sensor. Thus, any antinoise "heard" by the primary sensor does not find its way into succeeding antinoise outputs causing an unstable control loop.

11.7   Vibrations and Buildings

Another acoustical system to consider is that of vibrating equipment, its vibration isolators, and the building structure that supports the equipment. The pri-

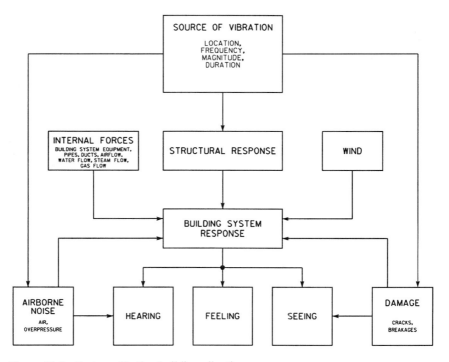

Figure 11.5   Factors affecting building vibration.

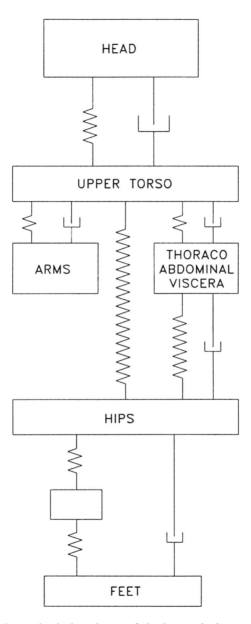

Figure 11.6 Simple mechanical analogue of the human body—a complex resonator. Bent knees and slouching shoulders damp out vibrations received via the feet. The response of body to other vibrations depends on direction, force, intensity, and distribution of vibration.

mary design goal is to mount the equipment on resilient isolators that are attached to very stiff, massive parts of the building structure.

Vibration in a building originates from both outside and inside the building. Sources outside a building include blasting, operations, road traffic, overhead aircraft, underground railways, earth movements, and weather conditions. Sources inside a building include doors closing, foot traffic, moving machinery, elevators, building HVAC and electrical systems, and other building services. Vibration is an omnipresent, integral part of the building environment. The effects of the vibration on building occupants depend on whether it is perceived by those persons and factors related to the building, the location of the building, the activities of the occupants in the building, and the perceived source and magnitude of the vibration. Factors influencing the acceptability of building vibration are presented in Figure 11.5.

Whatever the source of vibration, a person will perceive its effect by hearing it, seeing it, or feeling it. It is the combination of these perceptions that will determine human response. Because of this and the nature of vibration sources and building responses, it is convenient to consider building vibration in two categories—low-frequency vibrations less than 1 Hz and high-frequency vibrations of 1 to 80 Hz.

Human response to vibration depends on the vibration of the body. The main vibrational characteristics are vibration level, frequency, axis (and area of the body), and exposure time. See Figure 11.6 for a simple mechanical analogue of the human body, considered as a vibrating system. Tall buildings always oscillate at their natural frequency, but the deflection is small and the motion undetectable. In general, short buildings have a higher natural frequency of vibration than taller ones. However, strong wind forces energize the oscillation and increase the horizontal deflection, speed, and accelerations of the structure.

Higher frequency vibrations in buildings are caused by building system machinery, elevators, foot traffic, fans, pumps, and other process equipment. Further, the steel structures of modern buildings are good transmitters of high-frequency vibrations. Vibration does not reach us through the air, but audible sound does. A variety of insulations are barriers to sound transmission to a human space and reduce reflections within a human space. This art is part of the field of acoustics. One of the problems of mechanical design is to prevent structural parts from conducting sound, bypassing insulation.

12

Lightning, Electrostatic Discharge, and Buildings

12.1 Introduction

What causes lightning? Striking 100 times a second somewhere on the planet and packing temperatures up to 50,000 °F, lightning is both a common and a fearsome occurrence. Atmospheric electrical discharges known as *lightning* or *thunderbolts* (from cloud to cloud or cloud to ground) have captured the imagination and fear of the human race since ancient times. The ancient Greeks believed that lightning was Zeus' tool to punish human misbehavior or to demonstrate his anger.

Meteorologists recognize several types of clouds. When hot, humid air rises, it forms cumulus clouds. Large cumulus clouds rising to the top of the troposphere expand horizontally and form cumulonimbus clouds. These are the familiar "thunderheads" of the summer. Typically, electrical flashes are found in cumulonimbus clouds, but they may also occur in nimbostratus clouds (thick dark clouds at 2 to 12 km altitude covering the entire sky), in snowstorms and dust storms, and even in the gases of an active volcano.

Thunderstorms occur when a cloud becomes electrically charged. This can happen when ice crystals, water droplets, and other particles collide within the rising and falling air currents in clouds, producing electricity. The atmosphere usually works as an insulator to prevent this electricity from escaping. However, when the electricity stored in the thundercloud reaches a certain level the insulation effect breaks down and allows the instant formation of an enormous electrical current known as lightning.

Ball lightning is one of the most bizarre of all atmospheric phenomena. While ordinary lightning comes down in bolts from the sky, ball lightning is a fireball with a tail that moves close to the ground. Blazing with the brightness of several hundred light bulbs, ball lightning ranges in size from 6 inches to 2 feet in diameter and in shape from spherical to oval to wispy. Its color is usually white, sometimes with an orange or blue cast. But as varied as fireballs are in their appearance, they differ much more widely in behavior. Following paths that are straight or wildly curved, fireballs can float lazily over the ground for minutes or race by at thousands of miles per hour. They typically follow telephone or power lines for a few seconds, then vanish. According to some reports, fireballs also enter homes. Slipping through doors or windows, or diving down chimneys, they dart about furiously, then make a quick exit, leaving scorch marks in their wakes. Fireballs tend to occur in places where the air is stagnant, as over marshes and valleys. In cities, they usually form near high-tension wires, telephone lines, and the corners of metal buildings and towers. Nearly half of all reported fireballs appear inside buildings, with the fireball usually entering through a door or window.

People have reported seeing fireballs for centuries. But because ball lightning is so rare—and so fleeting when it occurs—it was not until the mid-twentieth century that scientists began studying it seriously. Three theories relate fireballs to lightning strikes. In the electricity theory, lightning strips hydrogen atoms from airborne water molecules. The hydrogen bonds with molecules containing carbon to form a ball of hydrocarbon molecules that emit light as they release excess energy. In the aerosol plasma theory, electrically charged particles called aerosols form a sphere. When lightning strikes, the sphere becomes energized and glows. In the electromagnetic wave theory, static electricity builds up in clouds and trees, generating electromagnetic waves that bounce off the ground. Where the waves meet, electromagnetic energy exits the air, forming a lightning ball.

12.2   Mechanisms and Characteristics of Lightning

The real cause of lightning is separation and accumulation of electrical charges in clouds via certain microphysical and macrophysical phenomena. To explain these phenomena initially precipitation and convection theory was developed, but today's most complete theory for lightning phenomena has established the fact that the structure of a thundercloud is *tripolar* (see Figure 12.1). This is the so-called charge-reversal hypothesis which states that when graupel particles (precipitation consisting of densely packed balls of snow or snow pellets) collide with ice crystals, the charge is transferred to a graupel particle. The polarity of the charge is dependent on the temperature.

At temperatures above a certain value, called the *charge-reversal temperature,* the transferred charge is positive. The exact value of the charge-reversal temperature is being debated, but it is believed to be around -15 °C. Considering that the temperature of the atmosphere is -15 °C (5 °F) at an approximate altitude of 6 km (3.75 miles), this means that owing to collisions of large snow pellets and ice crystals, the thundercloud will be, on aggregate, negatively charged for alti-

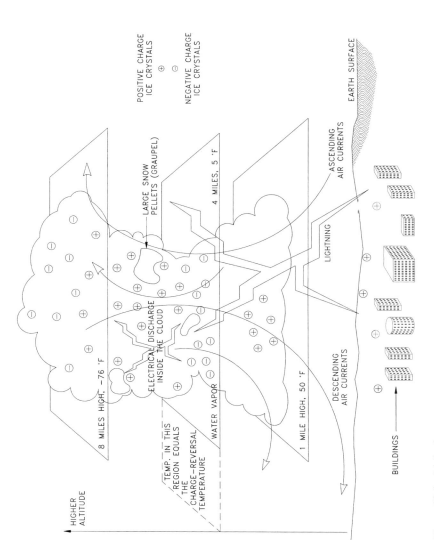

POSITIVE CHARGE
ICE CRYSTALS

⊕

NEGATIVE CHARGE
ICE CRYSTALS

⊖

EARTH SURFACE

LARGE SNOW
PELLETS (GRAUPEL)

4 MILES, 5 °F

ASCENDING
AIR CURRENTS

LIGHTNING

8 MILES HIGH, −76 °F

ELECTRICAL DISCHARGE
INSIDE THE CLOUD

TEMP. IN THIS
REGION EQUALS
THE
CHARGE-REVERSAL
TEMPERATURE

WATER VAPOR

1 MILE HIGH, 50 °F

DESCENDING
AIR CURRENTS

HIGHER
ALTITUDE

BUILDINGS

Figure 12.1 Lightning.

167

tudes about 6 km and positively charged below 6 km. This hypothesis has been verified in the laboratory and explains the levels of negative and positive charges in a thundercloud. Yet the exact microphysics of this phenomenon are practically unknown.

Mechanisms of Lightning: Lightning begins whenever the charge accumulation in a thundercloud is such that the electric field between charge centers inside the cloud or between cloud and earth is very high. For building engineering purposes, only cloud-to-earth lightning strokes (ground flashes) are of importance.

Typically, cloud to earth lightning stroke involves three stages. In the first stage, the high electric field intensity may generate local ionization and electric discharges known as *pilot streamers.* A pilot streamer is followed by the so-called *stepped leader.* The stepped leader is a sequence of electric discharges that are luminous; they propagate with a speed approximately 15% to 20% of the speed of light, and they are discrete, progressing approximately 50 meters at a time. The time between *steps* is a few microseconds to several tens of microseconds. The stepped leader will eventually reach the surface of the earth and will strike an object on the earth. However, where it will strike is not determined until the stepped leader is within a *striking distance* from the object.

The second stage initiates when the stepped leader reaches an object on the earth or meets an upward moving stepped leader. Specifically, a high-intensity discharge occurs through the channel established by the stepped leader. This discharge is extremely luminous and therefore visible. It propagates with a speed of about 10% to 50% of the speed of light. The return stroke carries an electric current of anywhere from few thousands of amperes to 200 thousands of amperes. The current magnitude rises fast, within 1 to 10 μs to the peak value, and then decreases rapidly. The discharge is known as the *return stroke* or simply the *lightning stroke.* The return stroke transfers a substantial amount of positive charge from the earth to the cloud and specifically to the charge center where the lightning originated. This transfer results in a significant lowering of the potential of the charge center. This phenomenon initiates the third stage of lightning.

In the third stage discharges may occur from other charge centers within the thundercloud to the depleted charge center because of the increased potential difference between them. This discharge will trigger another stroke between cloud and ground through the already established conductive channel with the first stroke. This process may be repeated several times depending on the electrification status of the thundercloud, resulting in multiple strokes.

Characteristics of Lightning Strokes: The parameters of lightning ground strokes are very important in the design of protection schemes against lightning. The most important parameters are

- Voltage
- Electric current

- Waveform
- Frequency of occurrence.

The voltage between a thundercloud and earth prior to a ground stroke has been estimated to be from 10 MV to 1000 MV. For design work, however, the building protection engineer is interested in the voltage appearing on the stricken power apparatus. This voltage will be equal to the product of the impedance times the stroke current. It is generally accepted that the ground stroke current is independent from the terminating impedance. The reason is that the terminating impedance is much lower than the resistance of the lightning discharge channel, which is on the order of few thousand ohms. Thus a ground stroke is normally considered as an ideal current source at the point of strike. The crest of the stroke electric current can vary over a wide range: 1 to 200 kA.

12.3 Building System Lightning Exposure and Protection

Building electric power systems are exposed to weather and therefore they are subjected to lightning strikes which result in overvoltages. Lightning overvoltages are generated by direct lightning strikes on a power system apparatus or indirect strikes to nearby objects, from which subsequent overvoltage is transferred to the system via inductive, capacitive, and conductive coupling.

Lightning overvoltages are independent of system voltage but depend on system impedances. For example, a direct lightning hit to a phase conductor of an overhead transmission line will generate an overvoltage proportional to the characteristic impedance of the line and proportional to the current magnitude of the lightning stroke. This overvoltage may be several million volts. It is a practical and economical impossibility to insulate distribution or lower-kilovolt-level transmission lines (i.e., 230 kV and below) to withstand this type of overvoltage. An integrated design procedure is applied to minimize the effects of lightning, which involves among other things: (1) shielding of lines and equipment, (2) effective grounding, and (3) application of protective devices (surge arresters). The presence of the shielding system ensures that lightning strike, which otherwise will discharge to a line conductor, will terminate on a wire-shield, air terminal, etc., that is electrically connected to the grounding system. A well-designed ground system will divert the majority of the lightning overvoltages.

Building distribution circuits are typically not insulated to withstand direct lightning strokes. As a result, direct strikes will cause a flashover. Direct strikes on building distribution lines are not frequent since the poles are not as high and therefore shielded from trees and structures. On the other hand, distribution lines may be vulnerable to overvoltages resulting from lightning strokes to nearby trees, ground, or other objects. Lightning strokes to nearby trees, ground, or other objects can result in voltage surges into the power system through coupling. The coupling can be conductive (through the conductive soil and the power system grounding structures), inductive, or capacitive. In a typical situation, all the coupling mechanisms may be present, resulting in a voltage surge to the power system. These voltages are called *induced voltage surges* and are generally much lower than those occurring after a direct strike. Specifically, they rarely exceed 400 kV.

It is therefore necessary to insulate distribution lines to withstand these surges. This translates into the requirement of a 300 kV basic lightning impulse insulation level (BIL) for distribution lines. In addition, power apparatus, connected to distribution lines and with BIL lower than the induced voltage surges, such as distribution transformers (typically the BIL of distribution transformers is 100 kV) must be protected. It is practical to protect transformers with surge arresters of appropriate ratings. Higher-kilovolt-level lines (i.e., 69 kV and above) have sufficient insulation withstand so that induced lightning voltages do not present the risk of flashover.

By far, lightning overvoltages on building power systems are the most stressful. Since these systems are interconnected to utility power systems, disturbances on the utility power system will be transmitted to them. These systems are also subjected to direct lightning. Therefore, these systems also require shielding against lightning.

Typically, a grounding system will be installed as well as a lightning protection system (shielding) to divert any direct lightning strokes to ground. This grounding system is referred to as *external grounding* to distinguish it from the so-called internal grounding, which refers to the grounding system of various types of equipment in the facility. The external grounding systems of industrial/commercial power systems are interconnected to the power system, as shown in Figures 12.2 and 12.3. The lightning protection system is basically a shielding system designed to route the lightning surges into the external grounding system for the purpose of minimizing potential differences within the facility. A system like this is subjected to lightning overvoltages which may enter from a number of points such as

- Air terminals
- Communication towers (if present)
- Power system grounding
- Fences.

An integrated design of the building external grounding system, lightning protection system, and internal grounding system can provide a system that is hardened against lightning and other sources of overvoltages.

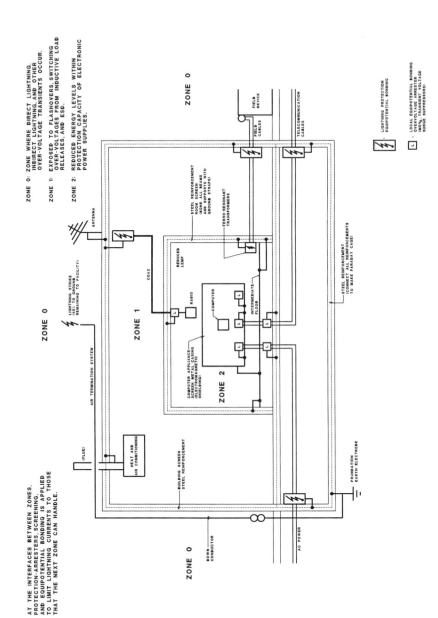

Figure 12.2 Zoned electromagnetic shielding and lightning protection diagram.

171

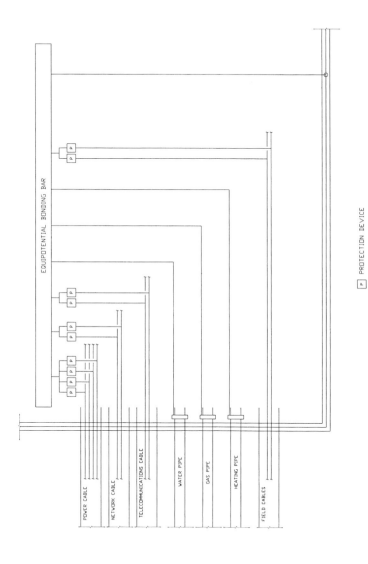

P PROTECTION DEVICE

Figure 12.3 Lightning and transient protection facility grounding diagram. Grounding of a facility goes a lot further than just the electrical systems. A full treatment of lightning and transient protection includes a grounding or equipotential system to shunt the high voltage and currents to ground potential in all zones. The total treatment would include current carrying conductors from lightning rods, located on the structure to rebar and metal inlay within the concrete, to zone protection conductors far within the structure.

12.4   Electrostatic Discharge (ESD)—An Introduction

To the average person the words "static electricity" can mean either a noise in the radio receiver that interferes with good reception or the electric shock experienced when touching a metal object after walking across a carpeted floor or sliding across the plastic seat cover of furnishings inside the building. Some people also have experienced mysterious crackling noises and a tendency for some of their clothing to cling or stick tightly together when wool, silk, or synthetic fiber garments are worn. Nearly everyone recognizes that this phenomenon occurs mainly when the atmosphere is very dry.

The word "electricity" is derived from the ancient Greek work *elektron,* meaning amber, for it was with this substance that the phenomenon of electrification was first observed. When the properties of flowing electricity were discovered, the word "static" came into use as a means of distinguishing the old from the new. The implication that such electricity is always at rest is erroneous; it is when it ceases to rest that it causes the most concern.

For the sake of simplicity, one might imagine electricity to be a weightless and indestructible fluid that can move freely through some substances, such as metals, which are called "conductors," but can flow with difficulty or not at all through or over the surface of a class of substances called "nonconductors" or "insulators." This latter group includes gases, glass, rubber, amber, resin, sulfur, paraffin, and most dry petroleum oils and many plastic materials.

When electricity is present on the surface of a nonconductive body, where it is trapped or prevented from escaping, it is called *static electricity.* Electricity on a conducting body that is in contact only with nonconductors is also prevented from escaping and is therefore nonmobile or "static." In either case, the body on which this electricity is evident is said to be "charged." The charge may be either positive (+) or negative (−). At one time it was thought that the two charges were two kinds of electricity and that in a neutral (uncharged) body they were present in exactly equal amounts. Now it is known that there is actually only one kind of electricity. It is true, however, that in a neutral or uncharged body the two entities are present in exactly equal amounts.

Work is required to separate positive and negative charges. Electricity, therefore, is sometimes referred to as a form of energy produced by expenditure of energy in some other form, such as mechanical, chemical, or thermal. Likewise, when electrical energy is expended, its equivalent appears in one of these other forms. These entities are components of all atoms, the outer electrons (−) and the inner (nuclear) protons (+). Electrons are free to move from one molecule to another in conductors but the proton, in the nucleus of the atom, cannot move appreciably unless the atom moves. Therefore, in solids, only the electrons are mobile; in gases and liquids, both are free to move. Curiously, a surface that has an excess or deficiency of one electron in every 100,000 atoms is very strongly charged.

The stable structure of the atom shows that unlike charges attract and, conversely, like charges repel. It follows that a separated charge will be self-repellent and will reside only on the surface of a charged body. If the body were a perfect insulator, the charge would remain indefinitely. However, there are no perfect insulators, and isolated charges soon leak away to join their counterparts and thus bring about neutralization, the normal state.

Static electricity then is the set of phenomena associated with the appearance of an electric charge on the surface of an imperfect insulator. It is "liberated" or usually made alive by the expenditure of mechanical work since electricity cannot be created. The charge on the surface of an insulator can thus attract or "bound" an equal and opposite charge on the nearest surface of any conducting body close to it. If the conducting body is now moved away from the originally charged body, the bound charge is now freed and will redistribute itself over the whole surface of the conducting body. In turn, it can be released in the form of a spark.

12.5   Electrostatic Charge—Charging Mechanism

Whenever two substances of different composition are brought into contact, one of the substances will surrender some of the electrons from its atoms to the other along the contact surface. Although the total (net) charge upon the two substances remains unchanged (and may be zero), the redistribution of charge resulting from this transfer of electrons results in the formation of an "electrical double layer" along the contact surface. One substance will have an increased abundance of electrons (and be negatively charged) while the other will be somewhat depleted of electrons (and be positively charged). Since these equal and opposite charges are strongly attracted to each other, they remain intimately related to the opposing surfaces and are not externally sensible so long as the surfaces remain in contact.

If the substances are nonconductive and are pulled apart, however, much of the charge disparity will remain with the individual substances, resulting in one being charged positively and one negatively. This charging mechanism is referred as "contact/separation or frictional charging." It can be enhanced by increased speed of separation, by lowered conductivity of substances, and by increased disparity in work function of the substances.

The surface of a substance that is subjected to bombardment by an ion shower (such as one originating at a corona point) will become charged by attachment of the ions or by surrender of charge to the surface by the ions. Charging by this mechanism is referred to as "bombardment charging."

When an uncharged object is brought into contact with another object that is charged, some charge will be transferred to the previously uncharged object. This charging mechanism is called "contact charging."

ESD is a charge-driven phenomenon. At any time, the net charge on a body is the difference between charge generated and charge dissipated.

$$Q_{net} = Q_{generated} - Q_{dissipated}$$

In many situations, induced charges are far more dangerous than the initially separated ones upon which they are dependent. Since a spark from the surface of an insulator can release a charge from only a small area, all the charge on the conducting body can be released in a single spark. In effect, a metal plate in close proximity to a charged surface can be considered one plate of a capacitor, and its ability to store energy is described as its capacitance. When a potential difference is applied between the two plates of a capacitor, electricity can be stored. In some instances one of the plates is the earth, the insulating medium is the air, and the other plate is some body or object insulated from the earth to which the charge has been transferred by induction or otherwise. Then a conducting path is made available, and the stored energy is released (the capacitor is discharged), possibly producing a spark. The energy so stored and released by the spark is related to the capacitance (C) and the voltage (V) in accordance with the following:

$$Energy = \frac{C(V)^2}{2}$$

If the object close to the highly charged nonconductor is itself a nonconductor, it will be polarized; that is, its constituent molecules will be oriented to some degree in the direction of the lines of force since their electrons have no true migratory freedom. Because of their polarizable nature, insulators and nonconductors are often called dielectrics. Their presence as separating media enhance the accumulation of charge. This is the reason dry area enhances accumulation and Electrostatic Discharge (ESD).

A person walking on an isolated floor covering can charge up to 10 kV. The typical capacitance to earth is 100 pF. When a person is close enough to a grounded structure (a computer, metal door, building steel, etc.), an ESD or arc is formed. Using the above figures, the charge the person carries is

$$Q = CV = (100 \times 10^{-12}) (10 \times 10^3) = 1 \text{ microcoulomb}$$

The current flowing is limited by the surge resistance of the person and varies from person to person, but it is significant enough to cause damage to electrical or electronic building system equipment.

12.6 ESD Solutions

One solution is to maintain a humid environment to reduce the dielectric resistance or use ionizers to dissipate or neutralize charge. Second, place a metal barrier between the person and building system electronic components. This doesn't make the problem go away but does protect the equipment. To solve the problem as well as prevent charge injection, it is necessary for the metal barrier either to completely surround the entire system (including cables), or to connect it to earth ground. An earth connection will bleed off the metal barrier charge and thus eliminate the electrostatic field, as well as prevent charge accumulation. In some cases, ESD can be reduced by very inexpensive means: use of static free carpeting, static sensitive building materials, and humidity control. ESD electromagnetic radiation usually can be prevented from causing malfunctions by use of sound maintenance practices. Replace missing items such as screws, shields, ground straps, and so on. Some degree of ESD potential control can also be obtained by momentarily grounding the human body upon entry into protection zone. This requires good coupling between charged sources associated with the human body (clothing and shoes) at the time of initial grounding and maintaining this coupling as the human body enters protected zone.

13

Electromagnetic Waves, Noise, and System Susceptibility

13.1 Introduction

To understand electromagnetic interference (EMI), electromagnetic compatibility (EMC), electromagnetic shielding (EMS), and their relation to building system sensitive equipment, it is important to understand how electromagnetic waves are created, their characteristics, and how they travel.

Any time an electric current varies its speed or course, it generates electromagnetic waves—which are nothing more than fluctuations of electrical and magnetic forces. Electric and magnetic fields are physical phenomena: invisible lines of force that occur whenever electrical power is used. Electric and magnetic fields are produced in energized conductors, electrical office equipment, internal wiring, and other items.

Electric and Magnetic Fields: Electric and magnetic fields are decoupled below radiofrequencies and are considered separate fields at low frequencies.

Electromagnetic Fields: Electric and magnetic fields at radiofrequencies and above are coupled together and referred to as electromagnetic fields.

The energy of an electromagnetic wave is related to its wavelength—the distance from one wave crest to the next. The shorter the wavelength, the more energetic the wave. In order of decreasing wavelength, electromagnetic waves include radio waves, infrared rays, visible light, ultraviolet light, x-rays, and gamma rays. Gamma rays are a mere one hundred-billionth of a meter or 3.3 hundred-billionths of a foot long, while radio waves can be a few miles long. As electromagnetic waves spread outward at the speed of light, their electric and magnetic fields radiate at right angles to each other and to the direction of wave flow (see Figure 13.1).

All electromagnetic radiation is transported exactly the same way through empty space. The atmosphere behaves almost like empty space except that parti-

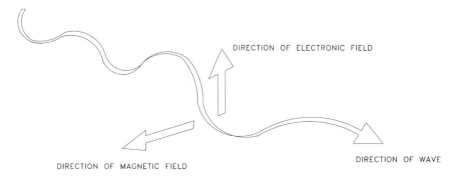

Figure 13.1 Electromagnetic Radiation.

cles and vapors selectively absorb radiation differently at different wavelengths. The electric field is a function of *only* the electric system voltage level. The higher the voltage, the stronger the electric field will be. The unit of measurement of the electric field is volts per meter. At higher levels of the spectrum it is electron volt/meter (eV/m).

The strength of the electric field decreases as an approximate function of the inverse square of the distance from the equipment or conductor containing the voltage. Thus, the strength of the electric field decreases rapidly as the distance from the source increases. Since a magnetic field is a function of moving charges (current), the magnet field can be looked on as closed loops of force lines such as ripples from a pebble dropping in the middle of a pond. With alternating current, the ripples would reverse with each change in current flow. The magnetic field is a direct function of the amount of electric current flowing in a piece of equipment or conductor. The larger the amount of current, the stronger the magnetic field will be. The strength of the magnetic field around a source decreases as an approximate function of the inverse cube of the distance from the electrical equipment carrying current. Thus, the strength of the magnetic field also decreases rapidly as the distance from the source increases.

13.2   Geomagnetic Fields and Buildings

A naturally occurring magnetic field exists and surrounds the earth. Humans and animals were conceived, nurtured, and grew under the influence of magnetic fields: Building systems operate in the company of natural fields. The SI unit of magnetic-field intensity is the tesla (T), equal to 1 newton/ampere meter. Other units in common use in geophysics are the gauss (G), equal to 10^{-4} T, and the ganna (y), equal to 10^{-5} G, or 1 nanotesla (nT). The magnetic north and south poles are defined on the basis of the compass needle—the end of the needle that points toward geographic north is called the *north magnetic pole,* and the opposite pole of the needle is called the *south magnetic pole.* The north pole of a compass needle points north because it is attracted by the earth's north magnetic pole, located in eastern Canada (77′18″N.101 48′W). However, a north magnetic pole attracts a south magnetic pole, and vice versa.

The intensity of the magnetic field of the earth at sea level ranges from 0.25 G at the magnetic equator to 0.60 G at the magnetic poles. By comparison, that of a small pocket magnet may range around 100 G.

The origin and dynamics of the earth's magnetic field still are not clear. The electron flow needed to support the magnetic field at its present intensity is about 44.10^9 amperes if it is in the form of a toroidal current in the outer core. In order to maintain the flow, an energy source is needed. Heat is the most obvious choice, but mechanical torques related to the precessional motions of the earth's axis have also been suggested. The trick is to transform heat and/or mechanical energy into an electron flow. It is obvious that the earth's rotation has something to do with the orientation of the magnetic field. There are a lot of free electrons in both the inner core and the outer core; there is a temperature gradient; and there is convection, at least in the outer core. The system is highly complicated in terms of electrical properties. It is likely that there are electrons flowing in all directions, except that there may be a minute excess flowing in a direction related to the earth's rotation, that is either overtaking the earth (and producing a normal magnetic epoch) or lagging behind (and producing a reversed magnetic epoch).

Electromagnetic interaction is mediated by photons. Four basic points need to be remembered in order to understand magnetism. The first point is that a charged particle moving with *uniform* motions creates around itself a magnetic field owing to continuous emission and reabsorption of virtual photons; the second point is that a magnetic field forces a change of direction in a charged particle moving at an angle through it; the third point is that a charged particle moving with *accelerated* motion (except electrons within orbitals) produces an electromagnetic wave; the fourth point is that an electromagnetic wave accelerates a charged particle.

If electrons flow with uniform motion through a straight wire, a circular magnetic field is created around the wire that has a counterclockwise direction when viewed along the direction of motion of the electrons. Conversely, if the electrons move with uniform motion around a loop on a counterclockwise direction, a magnetic field is created along the axis of the loop that is directed along the line of sight when viewing the electron flow in the loop as counterclockwise.

13.3   Electromagnetic Compatibility (EMC)

There are three elements to any EMC system: the source of the EMI called emitters; the media through which it is transmitted, conducted, or radiated; and the receptors or susceptors. Susceptibility is the tendency of a piece of equipment or a system to respond to undesired electromagnetic energy which otherwise affects the system adversely owing to the received EMI. Therefore EMC can be achieved by reducing the EMI emissions levels in the source, blocking the propagation path of the EMI signals, or by making the receiver less susceptible to the received EMI signals.

Figure 13.2 shows some of the emitters outside and inside of buildings. The source of the EMI is primarily any system where the current or voltage changes rapidly. Generally, external electrical noise is from the following sources:

1. Ignition noise from gasoline emergency generator engines
2. Static electric-type discharge from high-voltage power lines
3. The endless collection of business automation equipment.

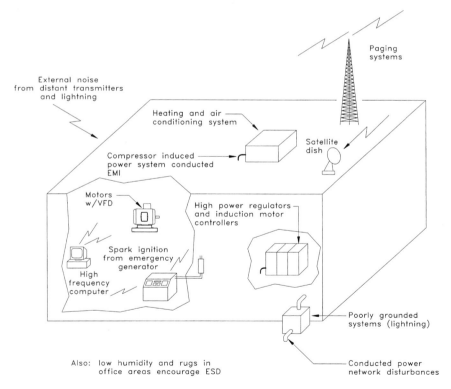

Figure 13.2 Interference can come from sources both inside and outside the building, and range from nearby radio and radar transmission to power line disturbances and lightning.

EMI sources can also be broadly divided into two categories, natural and man-made. Naturally caused EMI below 10 MHz is mainly due to atmospheric noise resulting from electrical storms. Above 10 MHz they are primarily a result of cosmic noise and solar radiation. A man-made noise common to all systems in buildings is power line hum. Building system operators used to determine if their telephones were working by noting the hum before attempting to use it. In addition there were the occasional pops and snaps from the power company doing some line switching. From the system point of view, it is necessary to have a way to determine how much noise each component will add to the system.

Power electronic circuits used in large building systems, by switching large amounts of current at high voltages, can generate electrical signals that affect other electronic systems. The electric field strength from these radiators is measured in units of volts per meter. When the field exceeds 100 V/m, EMI is almost certain to exist. When it is below 1 V/m, EMI is unlikely. These unwanted signals give rise to electromagnetic interference (EMI), also known as radiofrequency interference (RFI), since they occur at higher frequencies. The signals can be transmitted by radiation through space or by conduction along a cable (see Figure 13.3).

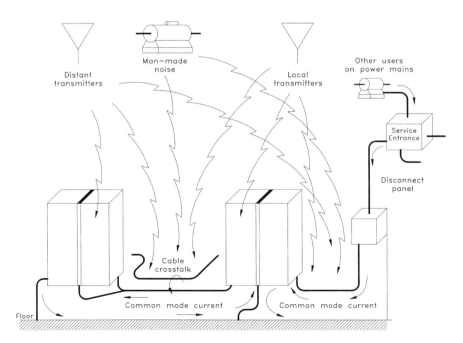

Figure 13.3   Electromagnetic interference can reach the victim through radiation or conduction.

Apart from emitting EMI, the control circuit of power systems can also be affected by EMI generated by its own power circuitry, by other circuits, or by natural phenomena. When this occurs the system is said to be susceptible to EMI. Any system that does not emit EMI above a given level, and is not affected by EMI, is stated to have achieved electromagnetic compatibility (EMC).

Naturally caused EMI through lightning can result in interference to ground or airborne equipment, and damage if a direct hit occurs. Peak currents can exceed 50 kA with rate of rise in the region of 100 kA/μs, *giving field strengths in nearby conductors of greater than 200 kV/m. The voltage induced in antenna systems, having a large physical length, is especially high and these are also prone to direct strikes.*

Man-made EMI can be intentional or unintentional. In both cases it is the variation of the voltage and current that produces EMI, whose magnitude depends on the value of the current, the length of the conductors, the rate of change of voltage and current, and the physical position of the conductors relative to each other and any earth plains. Examples of intentional EMI are buildings built near airfields which suffer most from EMI resulting from radar, where field strengths approach 200 V/m, with buildings giving only a low level of shielding.

Unintentional man-made interference is caused by sources such as switches, relays, motors, and fluorescent lights—equipment commonly found in buildings. The inrush current of transformers during turn-on is another source of interference, as is the rapid collapse of current in inductive elements, resulting in transient voltages. Integrated circuits also generate EMI owing to their high operating speeds and the close proximity of circuit elements on a silicon die, giving stray capacitive coupling elements.

Electromagnetic interference is generated in power circuits owing to rapid transitions and ringing. Oscillations can be damped by introducing resistance if the source of resonance is isolated. Harmonics generated by saturating transformers can be minimized by using high-permeability material for the core, although this would cause the device to operate at high flux densities and result in large inrush current. Electrostatic shielding is often used in transformers to minimize coupling between primary and secondary windings.

Interfering signals can often be bypassed by high-frequency capacitors, or metal screens used around circuitry to protect them from these signals. Twisted signal leads, or leads that are shielded, can be used to reduce coupling of interference signals. The collapse of flux in inductive circuits often results in high-voltage transients, causing interference in connecting circuitry. This is prevented by providing a path for the inductive current to flow, such as through a diode, zener diode, or voltage-dependent resistor.

EMI can be radiated through space, as electromagnetic waves, or it can be conducted as a current along a cable. Conduction can take the form of common-mode or differential-mode currents. For differential mode the currents are equal and opposite on the two wires and are caused primarily by other users on the same lines. These currents are mainly caused by coupling of radiated EMI to the power lines and by stray capacitive coupling to the body of the equipment (see Figure 13.3).

13.4   Electromagnetic Shielding (EMS)

The usual barrier to long waves is a surface of conductive material, either solid or mesh. The conductive surface is a short circuit to the electrical component of radiation. A sheet of high-permeability magnetic alloy is a barrier to both the magnetic component of radiation and to DC and AC magnetic fields. Refer to Figure 12.2.

Shielding Enclosures: The choice of which type of shielding system to use in a given building is a function of several factors. The first factor is performance. Only performance levels that are actually *needed* in terms of shielding effectiveness for each type of electromagnetic field and frequency range should be *specified.* The required shielding effectiveness is primarily based on the operational purpose of the facility. Hospital and secure communication facilities generally require shielding for different reasons. The operational purpose also determines the other physical factors required for the shielded enclosure such as appearance, HVAC, physical layout, and environmental requirements. All of these factors must be taken into account when selecting the type of shielding system for a given application.

A shielded enclosure is a complete structure with doors, vents, and other items. All penetrations to the shielded enclosure are treated to maintain shielding integrity. It may be a freestanding structure, such as the modular clamp-up structure, or a welded steel room built in place on the site. A more recent advent in shielding is the architectural shielding system, where metal foil or sheet metal is built into the walls, floor, and ceiling of existing or new construction.

The most common form of modular shielded enclosure is the clamp-up system using galvanized sheet metal bonded to both sides of a wood core and tied together by a framework made of plated steel. The galvanized panel system is the most common because it is readily available from a large number of suppliers and provides good performance over a broad frequency range. The magnetic field shielding effectiveness generally exceeds 100 dB for electric and plane-wave fields. The actual measured performance of these enclosures is a function of the installation workmanship.

As the size of the space to be shielded increases, the modular system becomes less cost effective. Finishing the shielded room with conventional wall coverings must be done very carefully to ensure that the integrity of the shield is not compromised. Painting of the shielding panels is not recommended since the paint forms capillaries under the strapping and negates the shielding properties of the system. The proof of shielding is in the performance test. This should be conducted by an independent testing service, and should be performed after all work that has an impact on the shielding is completed.

Water is an absolute enemy of shielding. Normal city water is loaded with minerals, and if spilled (coffee, etc.), the runoff can get under the strapping and form a deposit that would insulate the strap from the shielding panel and degrade the shielding effectiveness of the system. Typically a sealing material called "kobakoat" is used to seal the floor of shielding systems, preventing damage in the case of a sprinkler discharge. For critical mission facilities, it is recommended that a dry pipe system be installed. This system is activated by smoke detectors, with built-in delays, and alarms so that if any personnel are available, it can be verified that water is needed. A non-CFC gas extinguishing agent system is best for computer centers.

13.5   Electromagnetic Interference and Signal Protection

Signal and power conductors provide the simplest means of interconnecting the different elements in a system. It is not uncommon for these lines to be hundreds or even thousands of feet long. As these lines wind their way from source to destination, they often pass through areas with high electric and magnetic fields, which can severely distort the intended signals. Another threat to signal integrity is interference caused by ground loops and differences in ground potentials. Just keep in mind that, in addition to being a conductive path for noise, signal and power wiring can be a source and receptor of noise.

Crosstalk is a common noise problem caused by one system either radiating or coupling unwanted energy into another system. Refer to Figure 13.3. At lower frequencies, coupling is the major concern whereas stray radiation becomes dominant at higher frequencies. Just how undesirable crosstalk is depends on whether it is intelligible or not. If an interfering signal can be detected and has many of the same characteristics as the desired signal it is much more disturbing than an interfering signal of the same power but with different characteristics from the desired signal. An example is the crosstalk sometimes heard on the telephone. If the interference consists of a syllabic pattern, the hearer is considerably more distracted than if the interference is, say, a hum.

Coupled crosstalk is most often found in adjacent pairs of a multiple-pair cable. There will be a capacitance coupling between the wires owing to their proximity to each other and the dielectric between them. There is also the magnetic coupling due to a varying current flowing in one wire, causing a varying magnetic field that intersects the adjacent wire. The magnetic field intersection causes a voltage in the second wire. This magnetic coupling effect is the basic principle of a transformer.

The goal in signal protection from EMI is to minimize, divert, or eliminate one of the three elements necessary for an interference/noise problem. Usually, the element over which the system designer has the most control is the coupling path. Coupling can be capacitive or inductive, or conductive (through a common element).

Capacitive Coupling: Any piece of building equipment or wiring can develop an electric charge, or potential, which can be expressed as a voltage. If this charge changes, an electric field is generated that can couple capacitively to other equipment or wiring. This type of noise can be significant when a circuit or termination has a high impedance, because the noise voltage that's generated at the receiver is the product of the noise current and the receiver impedance.

An easy and effective way to minimize capacitively coupled interference is to use cable shielding. The shield is a Gaussian or equipotential surface on which electric fields can terminate and return to ground without affecting the internal conductors. Solid shields provide the best theoretical noise reduction solutions, but they're more difficult to manufacture and apply. Therefore, most cables are shielded with a braid for improved flexibility, strength, and ease of termination. Braided shields are less effective than solid shields because they provide only 60% to 98% coverage of the cable. Decreased effectiveness is more prevalent at high frequencies where the holes in the braid are large compared to a wavelength. For maximum shielding, reliability, and ease of use, cables with combined shields are available that use both a solid layer and a braided layer.

Shielding is effective against electric fields only if it provides a low-impedance path to ground. A floating shield provides no protection against interference. Grounding of shields can be a controversial subject because there are several ways to do it. *Pigtail* connections from a shield to ground have inductance, resulting in an impedance that increases with frequency. This type of connection will work at frequencies below 10 kHz, but will cause problems at higher frequencies. The use of short connections with large cross-sectional area minimizes the inductance of a pigtail, but the best connection is a 360° contact between the shield and connector. To complicate matters, some manufacturers of transmitters and receivers supply devices with cables whose shields have been internally connected and cannot be altered.

The correct place to connect an electrostatic shield is at the reference potential of the circuitry inside the shield. This point will vary depending upon whether the source and receiver are both grounded, or whether one or the other is floating. It is important to ground the shield at only one point to ensure that ground currents do not flow through the shield. In most applications, the shield ground should not be at a different voltage with respect to the reference potential of the circuitry. If it is, this voltage can be coupled to the shielded conductor through the capacitance.

The capacitance between conductors is inversely proportional to the distance between them. Therefore, another simple way to reduce capacitive coupling is to increase the distance between the victim cable and source cable. It is always a good idea to route *noisy* cables, such as power input wiring, motor control wiring, and relay control wiring, separate from *quiet* cables such as analog I/O lines, digital I/O lines, or LAN connections.

Magnetic Coupling: When a cable carries current, a magnetic field is generated. Magnetic coupling is much more difficult to reduce than capacitive coupling because magnetic fields can penetrate conductive shields. The amount of penetration is dependent on the frequency of the incident magnetic field, and can be related to the skin depth of the shield material. At one skin depth distance into a material, an incident wave will be attenuated by $1/e$ or 37% of its original value. In general, copper, aluminum, and steel are more effective shielding materials at higher frequencies, with steel providing about an order of magnitude increase in effectiveness over copper and aluminum.

The magnetic field can also be reduced by separating the source of the field from the receiving loop, or by reducing the length (loop area) of the conductors or by twisting the source wires if it is determined that the current producing the field flows in a wire pair. Twisting causes the magnetic fluxes from each wire to cancel each other out if equal currents flow in the wires, so the net field is ideally reduced to zero. If some of the return current were to flow through another path, such as a ground loop, a magnetic field would be generated by the imbalance in currents.

Twisting applies to both shielded and unshielded cables and to interference caused by shield currents or from other sources. Twisting the wires forces them close together, reducing the loop area and, therefore, reducing magnetic field generation and induced voltage. The effectiveness of twisted pair wire increases with the number of twists per unit length. In addition to reducing magnetic coupling, twisted pairs also act to reduce capacitive coupling. Each exhibits an equal capacitance to a noise source, causing equal and opposite charges to appear along the leads. This results in a net induced charge of zero and, ideally, no capacitively coupled noise. Twisted pairs are beneficial for use at frequencies below about 1 MHz. Above this limit, losses in the cable become a concern.

Some other common couplings encountered are: common-impedance coupling, in which two or more units or systems are connected to the same safety wire, ground grid, or plane at more than one place (multipoint grounding); common-mode and ground-loop coupling, in which radiated fields couple into ground loops that convert interference to undesited common-mode currents; and to differential-mode currents, differential-mode coupling; in which radiated fields penetrate signal and control cables to develop interfering voltages at the victim.

Unless all coupling paths are made sufficiently immune to EMI, electromagnetic compatibility may not result.

13.6 Grounding Systems of Buildings

In building systems environments, ground systems carry signal and power return currents, form references for electronic control analog and digital circuits, bleed off charge build-up, and protect people and equipment from faults and lightning. Because of these many requirements, ground is an elusive and often misunderstood concept.

The term *ground* implies that the soil we walk on is the place to which all currents and voltages levels of system are referred. This equipotential view of ground is not representative of practical grounding systems because two physically separated ground points are seldom at the same potential. Any current flow in a ground system can cause differences in potential. Lightning strikes or other transient events can generate hundreds to thousands of volts of potential difference.

It is not enough to design a system and leave grounding as a secondary issue. A ground system must be considered from the beginning in order for the system to work in the intended environment and pass interference, emissions, and safety requirements. There is no magic ground system that will work for all applications, so an understanding of the underlying principles is necessary for successful designs.

Proper grounding is dependent on many factors, such as the frequencies and impedances involved, the length of cabling required, and safety issues. When designing a ground or troubleshooting a ground problem, it is first necessary to determine where the current is flowing. If several kinds of grounds coexist, the current may not return by the assumed path.

To maintain high power quality a proper grounding and bonding is a must. The signal ground should have a low impedance to handle large signal currents, and this is usually done by making the ground plane large. The inductance is reduced by placing the signal current-carrying conductor close to its ground return. Single-point grounds are difficult to maintain at a low impedance and are not suitable for frequencies above about 10 MHz. Generally, above these frequencies multipoint grounds are used. However, care is now needed to prevent the occurrence of ground loops, which can generate fields that interfere with the signal.

The most desirable type of ground for low-frequency applications is the single-point ground (see Figure 13.4). Two examples are shown in Figure 13.5. Avoid the series connection, or daisy-chain, when sensitive equipment is involved, because return currents from all the equipment circuits flow through the common ground impedances linking the circuits. The ground potential of the equipment-1 circuit is determined not only by its return current through impedance Z_1, but also by the return currents from equipment-2 and -3 circuits through the same impedance. This effect, which is called common-impedance coupling, is a primary means of noise coupling. Equipment circuits were used in the example, but any grounding scheme follows the same guidelines. Beware of common-impedance coupling when two signal conductors share one ground.

The preferred ground is the parallel connection. It is usually more difficult and more costly to implement because of the amount of wiring involved. When choosing between these ground configurations, first determine the circuit common-

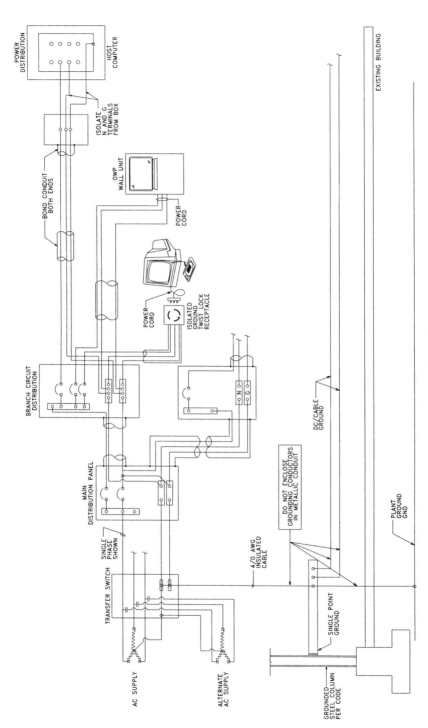

Figure 13.4 Single point ground (SPG) dedicated to control system wiring diagram.

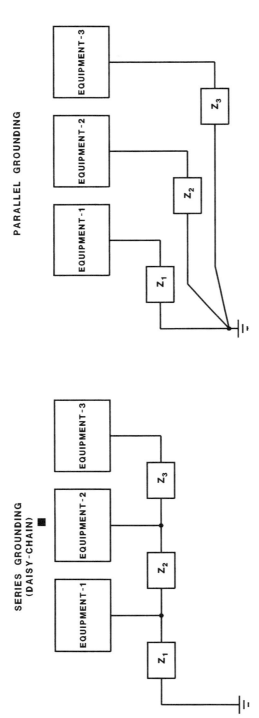

Figure 13.5 Single-point equipment grounding methods.

impedance coupled noise immunity. In practice, most systems will use a combination of both topologies.

13.6.1 High-Frequency Interference and Shield Grounding

The proliferation of electronics in building systems is generating an environment with increasing electromagnetic pollution. As the speeds of electronic circuits continue to increase, electromagnetic interference will present a continuing challenge to building systems engineers. Shielding and grounding concepts have remained constant throughout all the new developments in electronics, but now more than ever, it is important to understand these concepts and to implement them in circuits and systems from the initial design stages.

Single-point grounds for cable shields work well up to about 1 MHz depending on system size. At higher frequencies, parasitic capacitances provide sneak paths for shield currents that form ground loops. Also, standing waves present on a shield cause the impedance to vary along the shield. So even if there is a low-impedance connection to ground at one end of the shield, the impedance will go to infinity at points that are odd multiples of $1/4$ wavelength apart. For digital equipment cabling, or when the shield length exceeds $1/20$ wavelengths of the highest frequency or harmonic present, a cable shield is often grounded at both ends and possibly at several points in between, depending on cable length and frequencies present.

Multiple shield grounds may seem contradictory to previous statements, but the noise induced by ground loops is generally at lower frequencies and can be filtered out, preserving the high-frequency signals. Also, at high frequencies, the skin effect causes signal current to flow on the inside surface of the shield, and noise current to flow on the outside surface of the shield; thus, the shielding benefits of a triaxial cable are realized. Most systems have a hybrid ground system where the shield will have a single-point ground at low frequencies and multiple grounds at high frequencies.

13.6.2 Ground Loops

Ground loops exist in a system when there are multiple current return paths or multiple connections to *earth ground.* Current flowing in a ground loop generates a noise voltage in the circuit. No matter when you might think, buildings do have many ground loops, some of which may include cable shields (Figures 13.6 and 13.7). Here are a few causes associated with building systems:

- Both commercial power and UPS safety grounds entering the cabinet
- Cabinet grounded to building ground grid and/or steel and indirectly grounded to other ground points

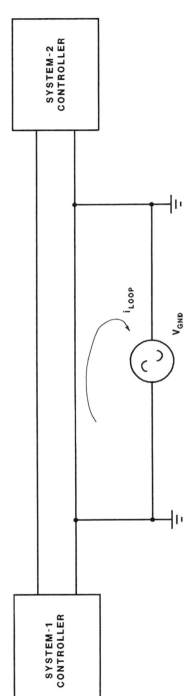

Figure 13.6 Ground loop between two systems.

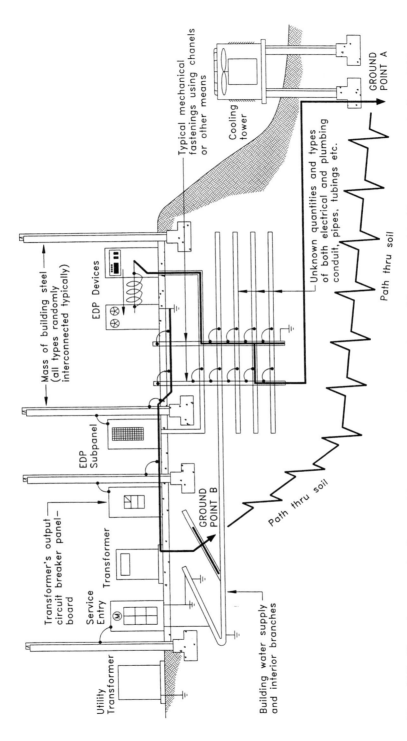

Figure 13.7 Shown is an example of a ground loop. Here ground point A may have a different potential with respect to ground point B. With some impedance between the two points, we have a potential driving a current around the loop.

- Internal dedicated analog reference (returns) grounded via long external runs to building ground
- Analog and logic returns connected together (often indirectly) and grounded to the cabinet
- Uncovered cable trays multipoint grounded to building steel.

The most obvious way to eliminate the loop is to break the loop connection between the equipment and ground. When this isn't possible, isolation of the two circuits is a universal way to break the loop. Isolation prevents ground loop currents from flowing and rejects ground voltage differences. A popular method of isolation involves the use of signal conditioners based on transformers or optical couplers. Signal conditioners often are rated to withstand transient events, thus providing a level of protection for the host system against harsh building system environments and costly damage in fault situations.

14

The Human Body and Building Synchronization

14.1 Introduction

The body organs are integrated to produce organ systems. Unlike most building systems, body systems are highly organized. To maintain this order, energy must be fed into the system. All living cells depend on the release and use of energy for the maintenance and operation of the living state, just like building systems. However, no building machine or body system can produce energy out of nothing; it can only change energy from one form to another.

Although all of life depends on the sun, only photosynthesizing creatures are capable of trapping light energy directly and, by using simple raw materials, synthesizing organic compounds; these organisms are called *producers*. In discussions of energy, the term "producer" may sometimes be misleading. Producers do not, indeed they cannot, generate energy out of nothing; rather, they convert light energy into chemical energy. Producers are, in fact, converters or transducers of energy, taking the radiant energy of the sun and chemical elements from the earth and transforming these into useful, energy-rich organic molecules. Because photosynthesizing organisms synthesize their own food, these producers are called "self-feeders."

Laws of energy transformations such as that of thermodynamics apply equally to buildings and living beings. The human body is an open thermodynamic system, that is, it must continually be supplied with energy to maintain itself. Energy supplied to the body in the form of food cannot be utilized with 100% efficiency for cellular work, and at every energy transformation there is a reduction in the usable energy of the system—usually in the form of heat. This is similar to heat dissipation in buildings, which is part of the energy unavailable for work to operate the system.

The body's energy storage involves breaking the energy-rich chemical bonds of foods through oxidation, and trapping a portion of released energy in the chemical bonds of high-energy phosphate compounds called ATP (adenosine triphosphate). This storage is similar to electrochemical storage in buildings. ATP can be likened to the "charged" form of the storage battery. Just as a battery runs down and becomes discharged when the chemicals in the electrodes no longer donate electrons, so too there is a "discharged" form of ATP called ADP (adenosine diphosphate). The link between ATP and ADP releases a great deal of energy, like the electron flow link between cathode and anode in a battery. Discharged ATP needs recharging, like batteries in buildings.

The human body is capable of using a variety of foods as sources of energy, roughly equivalent to building systems that could operate on gas, coal, wood, or fuel. The building system lacks such versatility, but humans do not. This is largely achieved because ATP is the single mediator (like currency in commercial systems) between the energy suppliers and the energy users, unlike buildings that have series of mediators that provide a direct usable form of energy such as electricity, steam, or domesticized gas. In the body ATP provides the driving force for most cellular activities and mechanical, chemical, osmotic, and electrical work whereas in buildings the driving force takes many forms such as chilled water, hot water, compressed air, power, steam, and gas. For this reason, a human cell is about 40% efficient as an energy trapping machine; by comparison modern power plants are only 30% efficient, with the overall system, taking into account source loss, 15% to 20% efficient.

14.2 The Body-Building Environment

Our bodies strive to maintain, at all costs, a nearly constant core temperature for our vital organs. This most protected zone takes thermal precedence over the less vital zone of our extremities, such as arms and legs; next down in priority are our fingers and toes. The most variable thermal zone of all is our skin surface. Similarly, buildings are frequently thermally zoned, and users (paralleling human blood flow) can retreat from—or advance to—the less protected zones as conditions demand.

To maintain the cores of our bodies within a narrow temperature range, we are always generating bodily heat and need to loose this internally produced heat to our environment. The rate at which we produce heat changes frequently, as does the environment's ability to accept or reject heat. To regulate our bodily heat loss, we have available three common layers between our body cores and our environment: the first skin, our own; the second skin, clothing; and the third skin, a building envelope.

Once the blood and water get our surplus heat from the organ system to the skin surface, we have four ways to pass it to the environment: *convection* (air molecules pass by our surface, absorbing heat); *conduction* (we touch cooler surfaces, and heat is transferred); *radiation* (when our skin surface is hotter than other surfaces "seen" but not touched, heat is radiated to these cooler surfaces); and *evaporation* (a liquid can evaporate only by removal of large amounts of heat from the surface it is leaving). The amount of heat we lose by each of these four methods depends on the interaction of our metabolism, our clothing, and our environment. As air and surface temperature approach our own body temperature we lose the options of convection, conduction, and radiation. Evaporation is essential, so access to dry, moving air is greatly appreciated.

As air and surface temperature fall, evaporation drops while convection, conduction, and particularly radiation increase. Usually clothing acts as an insulating layer and is particularly effective at retarding radiation, convection, and conduction. As air and surface temperatures fall well below our own, we adjust the second skin. Our second skin is just as likely as our third skin to be dominated by considerations of style more than of thermal regulation; we can't always count on clothing—or buildings with no environmental systems—to increase our comfort.

A positive definition of comfort is "a feeling of well-being." However, the more common experience of thermal comfort is a lack of discomfort—or being unconscious of how you are losing heat to your environment. The interaction between comfort and those environmental factors can be generally summarized in Figure 14.1. The "comfort zone" represents combinations of air temperature and relative humidity that most often produce comfort. Building systems can provide comfort zones inside buildings.

If our second skin does not provide comfort conditions then our first skin responds by internal activity. See Figure 14.2 to draw a parallel between the body and building air conditioning systems. When we are too cold, we begin to increase heat production through shivering and nondirected increases in metabolic rate. Shivering, incidentally, is uncoordinated muscular activity with no other clear purpose than generation of heat. For short periods shivering can raise resting heat production as much as fivefold; for periods as long as an hour bouts of shivering can give a two- or threefold increase. Piloerection ("goose bumps")—fluffing up the fur—is a bad joke in humans, but of great use to more ordinarily furry mammals to conserve body heat.

The opposite occurs when we are too hot; blood flow toward the skin surface increases (vasodilation), and the sweat glands greatly increase their secretion of water and salt to the skin surface. We're especially good at sweating, but not at all effective at panting. This increases heat loss by evaporation. The primary sensor in these thermoregulatory feedback loops is located at the base of the brain.

The presence of temperature sensors in the skin is obvious to us all, since we quite literally feel their activity; but do they play a role in thermoregulation? The answer is slightly peculiar. The sensors turn out to be much more important in the response to cold than in the response to heat. When a person is exposed to a

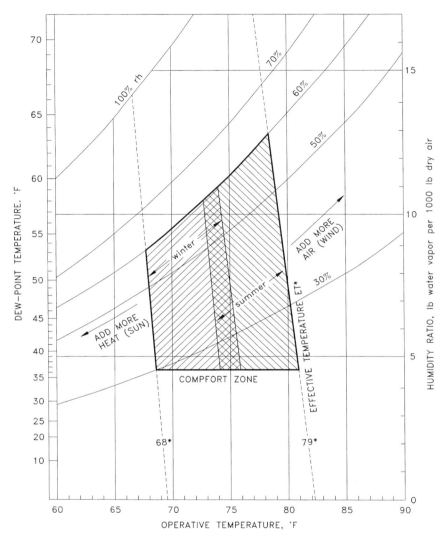

Figure 14.1　Standard temperature and humidity comfort zone. (Reprinted by permission of the American Society of Heating, Refrigerating and Air Conditioning Engineers, Atlanta, Georgia, from the *1993 ASHRAE Handbook—Fundamentals*)

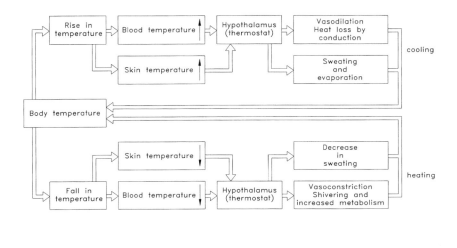

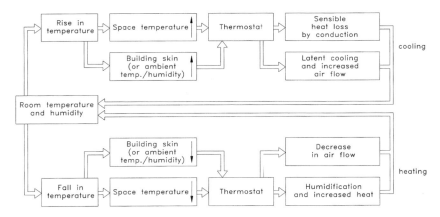

Figure 14.2 (Top) Feedback loops involved in the maintenance of constant body temperature. (Bottom) Feedback loops involved in the maintenance of constant room temperature.

heat load, skin temperature changes in a direction opposite that of sweating rate—the more you sweat (really sweat, evaporating the water), the cooler is your skin. Since sweating cools the skin, almost as soon as sweating starts the sensors in the skin would send quite an inappropriate message, saying that all is well. That message would arrive long before any significant cooling of the core could happen, and it would thus turn off the cooling devices prematurely. Why the thermostat is inside the building should now be obvious.

The human body uses its own anticipatory control to conserve energy, like building systems. For instance, if you begin exercising, then the necessary increases in breathing and circulation start *before* the oxygen and carbon dioxide concentrations in the blood have changed. Both diffusion and circulation take time, and the body anticipates the *lag*. With the increased neural traffic to and from the muscles, the brain, of course, knows that you're up to something. The system therefore has an additional input, one that we might call a "change" detector as opposed to the "error" detecting sensors within the loops. Anticipatory control is always secondary, even though it goes into action earlier, and its role is reduced when the feedback loops get properly aroused.

The longer a feedback system takes either to respond or to recognize that it has responded, the more important anticipatory control becomes. If you're thirsty, you drink water. If you kept drinking until the water was actually absorbed through the wall of the stomach, you'd have far exceeded that needed to replace the deficit underlying the thirst, and you'd then excrete a lot of water. In short, the system would overshoot badly and perhaps oscillate. Anticipatory control by sensing things such as stomach distention compensates for the relatively long lag of the fundamental, nutritional loops. These feedback loops are like PID (proportional, integral, and differential) control loops commonly applied to maintain the indoor environment of buildings within a comfort zone.

14.3 Electrodynamics of Humans and Buildings

To explain this a popular model is that of the self-exciting dynamo: A conductor rotating in a magnetic field creates an electron flow. If the flow passes through a coil, the coil itself creates a magnetic field that keeps the flow moving as long as the conductor keeps rotating. Well, the earth rotates. The core is conducting, but probably more here and less there because of temperature and/or compositional inhomogeneities. Although the ingredients for the creation of a magnetic field are there, the exact mechanism is still not known. While moving through this naturally occurring earth magnetic field, humans generate electric voltage within his body by induction. Human bodies compensate for externally induced voltages by the development of internal systems that generate electrical pulses that are high enough to overcome this interference generated externally.

The electric field is a function of *only* the electric system voltage level. The higher the voltage, the stronger the electric field will be. The unit of measurement of the electric field is volts per meter. The conductor does not have to have current flowing to generate an electric field. The conductor has to be energized to have an electric field present. An electric field is present in an electric toaster that is plugged into a wall receptacle yet not toasting anything, even if the switch is open. The electric fields of the human body are generated by movement of muscles. To evaluate and compare the effect of an external electric field on the human body, the magnitudes of some natural body electric fields are cited.

The heart produces electric fields on the surface of the chest that are on the order of 50 mV/m. In the area of the brain, fields of 1 V/m over an extremely small area are produced, with values of 1 mV/m over an area of a few centimeters. The differential voltage ("radical differential voltage"—chiropractic term) across the spinal cord that appears on the skin surface can be as high as 13.5 mV standing or as low as 5.5 mV sitting. To produce neural stimulation, the tissue fields are in the order of 1 to 100 V/m. Brain nerve cells typically produce 0.1 volt, 2 millisecond spike trains with repetition rates in the range of 1 to 100 Hz.

The human cell has two distinct electrical parts. The membrane of the cell acts like an insulator, and any 60-Hz induced current flows around the cell. The individual cells in the human body can have an electrical field at the surface of the cell up to 10 MV/m (10 million V/m). Through Faraday's law one should calculate the typical voltage induced by a magnetic field across a cell membrane to determine what body can or cannot tolerate. The levels of internal body voltages, currents, and fields are several magnitudes greater than those of externally induced electrical and magnetic fields to immune body from small level electrical jolts.

Hazardous Voltage and Currents: The human body is susceptible to direct application of electric current. The human body can withstand electrical abuse, but there is a limit to what the body can withstand. The upper end is electrocution. Teetering on the border is electroshock therapy. Safety clearances and strict code requirements for building systems set the "safe" limit.

To electrocute a person, the electric chair uses a 5-ampere current at 2640 V. Two thousand volts is usually sufficient to stop the heart, and an extra 640 V are added to compensate for persons with large body mass and any additional voltage drop. Five amperes is used as this amount will not burn the human body. The current is applied for two 1-minute periods with a 10-second delay between jolts. The time period is necessary to ensure death.

Electroshock therapy uses 350 to 800 V peak. The time of application varies from 2 milliseconds to 2 seconds with a period of 0.5 to 2 seconds between applications. The amount of power applied ranges from 59 to 100 J. One source equated the power to that of a 40-W lamp. The application of this amount of current to the brain modifies memory and personality.

In treating heart failure, it is not unusual to apply a momentary shock of 50 W/second or 400 to 500 W/second of direct current at a cycle of 6 seconds between shocks. This application, which is at the surface of the skin, will usually reset the heart firing cycle to halt ventricular flutter or fibrillation.

To avoid the above, the maximum amount of current that has been established as acceptable and safe for humans is 9 mA. Between 30 and 250 mA is the danger zone, resulting in cessation of breathing. Beginning at 75 mA, the heart can go into fibrillation. *Ground fault circuit interrupters* (GFCIs), which are used and required by the National Electrical Code, are set to trip at 5 mA.

14.3.1   Radiofrequency Waves and the Human Body

Radiofrequency waves are present in all modern buildings. This section points out some of the effects of radio waves on the human body. Just as the body absorbs infrared and light energy, which can affect thermal balance, it can also absorb other longer wavelength electromagnetic radiation. Visible light (wavelengths 0.4 to 0.7 μm) and infrared (wavelength 0.7 to 10 μm) are absorbed within 1 mm of the body surface. The heat of the absorbed radiation raises the skin temperature and, if sufficient, is detected by the skin's thermoreceptors, warning the person of the possible thermal danger. With increasing wavelength, the radiation penetrates deeper into the body. The energy can thus be deposited well beneath the skin's thermoreceptors, making the person less able or slower to detect and be warned of the radiation. Physiologically, these longer waves only heat the tissue and, because the heat may be deeper and less detectable, the maximum power density of such waves in occupied areas is regulated (see Figure 14.3).

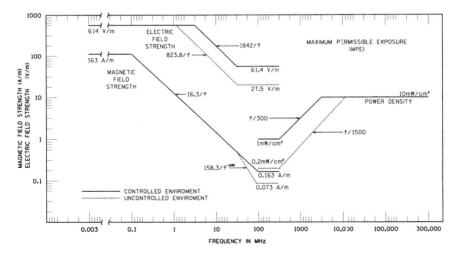

Figure 14.3   Maximum permissible levels of radiofrequency for human exposure. Note: Maximum permitted power densities are less than half of sensory threshold values. (Reprinted by permission of the American Society of Heating, Refrigerating and Air-Conditioning Engineers, Atlanta, Georgia, from the *1993 ASHRAE Handbook—Fundamentals*)

14.4   Human Body Systems versus Building Systems

The principal elements of building systems that can be compared with body systems are blood composition (analogous to air quality in building systems), blood pressure (analogous to pumping pressure in buildings), and body temperature control, analogous to building temperature control. These are regulated by adjusting other items such as cardiac (control) output, the resistance (friction) of the vessels (pipes) of the microcirculation, and the relative apportionment of blood (power) to the various organs. These parallels are depicted in a split graphic (Figure 14.4), half of which shows a building and the other half a human body.

14.4.1   Human Circulatory versus Building Hydraulic System

The human heart is about the size of a clenched fist. In a life span of 70 years it beats 2.5 billion times and pumps 40 million gallons of blood. The heart is indispensable to one's very existence. The heart pumps blood, oxygen, and food to every cell of the body, and by the return circuit of the bloodstream, cellular wastes and carbon dioxide are removed. See Figure 14.5 for a comparison between a human circulatory system arrangement and a common arrangement of a building hydraulic system. The human heart has an energy demand, and this varies with its work load, similar to the energy demand of a building hydraulic system. For example, the heart of an obese individual requires more energy to operate than that of a person of more normal girth. It has been estimated that for every pound of fat there are three extra miles of blood vessels that require pumping. Similarly large building systems require more pumping and hence more energy.

Both the mechanical integrity of the circulatory system and proper exchange across capillary walls require that the pressure of the circulating fluid be kept within reasonable limits. Average blood pressure is mainly determined by two factors—how hard the heart is pumping and the overall resistance of the microcirculatory vessels. A change of circumstance may lead to an increase in the resistance of some, to a decrease in others, and leave still others unchanged. In a building pumping system, a good flow control system works over a range of flows, of temperature, of pressure, of loads, and so forth. The main trick is to put the sensor where control needs to be most effective, where alterations in the controlled variable are least tolerable.

All too many of us regulate our blood pressures at levels too high for our own best interests. In many people with high blood pressure, because the arteries are stiffer, the body pressure sensor is fooled by less pressure on the receptors in the artery walls, so the system will "think" the pressure is lower than it actually is. Thus it will regulate at a higher pressure. Similarly too many buildings operate with oversized pumps ("just to be on the safe side"), and system control is made to regulate flow at a higher pressure.

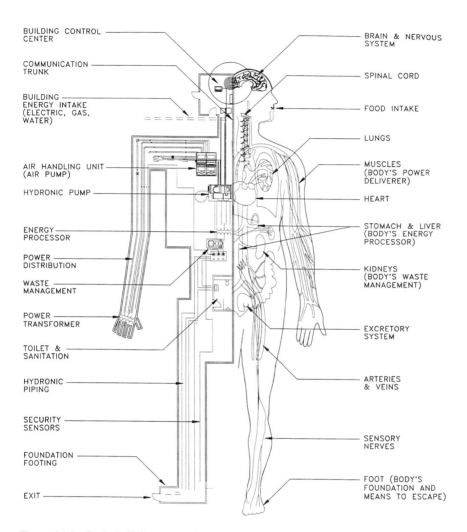

BUILDING CONTROL CENTER
COMMUNICATION TRUNK
BUILDING ENERGY INTAKE (ELECTRIC, GAS, WATER)
AIR HANDLING UNIT (AIR PUMP)
HYDRONIC PUMP
ENERGY PROCESSOR
POWER DISTRIBUTION
WASTE MANAGEMENT
POWER TRANSFORMER
TOILET & SANITATION
HYDRONIC PIPING
SECURITY SENSORS
FOUNDATION FOOTING
EXIT

BRAIN & NERVOUS SYSTEM
SPINAL CORD
FOOD INTAKE
LUNGS
MUSCLES (BODY'S POWER DELIVERER)
HEART
STOMACH & LIVER (BODY'S ENERGY PROCESSOR)
KIDNEYS (BODY'S WASTE MANAGEMENT)
EXCRETORY SYSTEM
ARTERIES & VEINS
SENSORY NERVES
FOOT (BODY'S FOUNDATION AND MEANS TO ESCAPE)

Figure 14.4   Body-building system integration.

In the human body the amount of heat that can be exchanged by conduction across moist interfaces depends on the area exposed; the respiratory surface can be increased tremendously by extensive folding of the surface. Similarly in air handling units the coil surface area is increased by folding more rows and adding fins. Once the air has passed across the respiratory surface, efficiency of distribution is increased greatly if the respiratory surface is amply supplied with blood. The same applies to flow in coils inside air handling units.

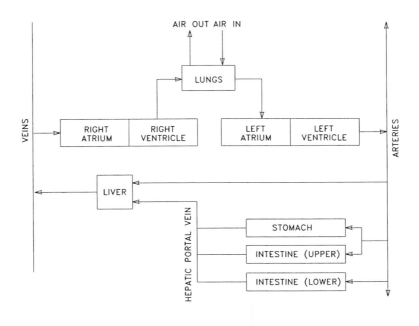

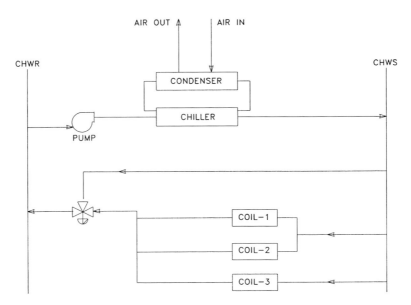

Figure 14.5 (Top) Human circulatory system. The circulatory connections of our liver and functionally adjacent organs, pointing out the way its portal system supplies it with venous blood from the intestines. (Bottom) Building chilled water system.

14.4.2   Human Respiratory versus Building Air System

The exchange of gases between the living organism and its environment is called *respiration.* Respiration in humans involves two related processes. The direct exchange of gases between environment and organism is called *breathing* or *ventilation;* this exchange in the body involves oxygen transfer from the respiratory system to the blood, which carries oxygen to every cell of the body and exchanges it for carbon dioxide. The carbon dioxide is carried back to the respiratory system, where it, in turn, is given up in exchange for more oxygen. In buildings "respiratory" systems, oxygen is transferred from the atmosphere to the air circulatory system which supplies oxygen to all internal spaces and exchanges it for carbon dioxide. The carbon dioxide is carried back to the air handler where it is exchanged for more fresh air. Although oxygen and carbon dioxide both cross the interface of the lungs, they move independently of each other. They are like total strangers who simultaneously touch the station platform, one boarding and the other leaving the same train. Building air handling units have similar separate chambers. See Figure 14.6.

The human air filtering and conditioning system has unique similarities with building air systems. Air first enters the respiratory tree through the nostrils, where projecting hairs filter out dust and debris. The air then enters the nasal cavity, which is divided by a septum and lined by a ciliated epithelium. The cilia distribute a mucus film that traps debris, and in conveyor-belt fashion this mucus sheet is moved toward the throat, where it is swallowed or spat out. Beneath the nasal epithelium is a rich network of blood vessels that helps to warm the air as it eddies about in the nasal cavity. Scrollike turbinate bones and air spaces in the skull (sinuses) form a complex labyrinth of passages that further assist in filtering, warming, and moisturizing the incoming air. Even our tears help moisturize the air we breathe, since tear ducts drain the continuously operating tear glands and empty directly into the nasal chamber.

Warmed, humidified, and scoured of dust and small particles, the air passes from the nasal cavity to the throat (pharynx), where the air and food passages cross. From the pharynx the air enters the larynx (voice box, or Adam's apple) through a slitlike opening, the glottis, which is guarded by an elastic flap, the epiglottis. The glottis in human body acts like damper in building air handling unit, the glottis remain open at all times except when we swallow similarly damper in air handling unit remains open at all times except when air is bypassed.

With each breath we take in 500 ml of air, but only 350 ml of that air actually reaches the alveoli and remainder 150 ml is trapped in the upper respiratory passages. Similarly in buildings all the ventilation air does not reach work area some air remains trapped in stagnant zones of the building. See Figure 14.7.

Human beings take 4 to 10 million breaths a year equivalent to a fan running at 15 revolutions per minute all the time. During the passage of air from the nose

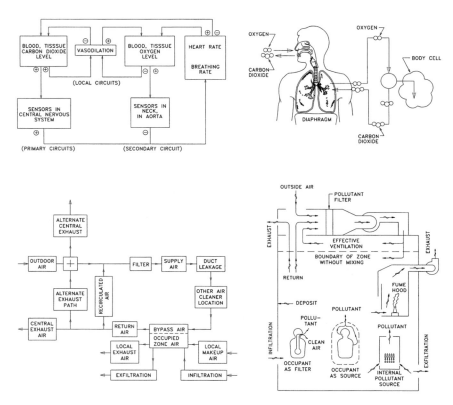

Figure 14.6 (Top) Body control system to control the level of oxygen (and carbon dioxide) in the blood. (Bottom) Building IAQ system to control the level of oxygen and other impurities inside the building.

to the lungs, any number of things may go wrong but nothing goes wrong and respiratory system function reliably. To learn how to increase reliability of building environmental systems a study of reliable human respiratory systems can provide useful tips. The respiratory center has a built-in safety factor with two feedback loops: one loop is intrinsic in the medulla and the other is extrinsic through the reflex system. This redundant control system is characteristic of many important body functions and is essential for a system that must operate without failure in it's lifetime.

How does the respiratory center "know" what the body's oxygen requirements are? At first glance it might seem sensible for the amount of oxygen in the blood to influence the respiratory center directly, so that when the blood oxygen level goes down, ventilation increases to compensate; however, this turns out not to be the case. Increased levels of carbon dioxide in indoor air rather than decreased

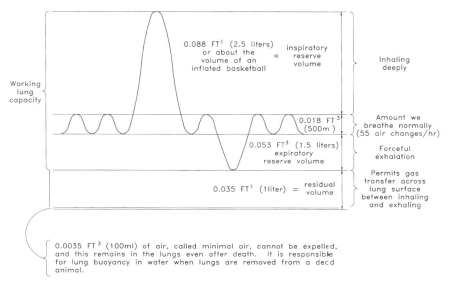

Figure 14.7 Human body air handling system flow control. The working lung capacity is about 0.0159 Ft³ (4.5 liters) as measured by deep breathing in and out, but in actual fact the total lung capacity is 0.194 Ft³ (5.5 liters), with 0.03531 Ft³ (1 liter) as residual air. To match air changes with breathing rate, a building system needs to provide 55 air changes/hour compared to normal 10 to 12 air changes/hour.

levels of oxygen stimulate the respiratory center to activity. Although the respiratory centers of the brain monitor carbon dioxide levels in the blood and thus regulate ventilation, there is an independent, or backup, system that registers changes in the amount of oxygen in the blood. The transfer of gases between the blood and other tissues of the body is called *internal respiration* and strictly speaking is a function not of the external respiratory system but of the circulatory system.

14.5 Human Body System Automation versus Building System Automation

The body receives messages from its own organs and from the external world. We perceive light, sound, odors, pressure, temperature, chemicals, and the like; we think; we move; we have unconscious thoughts and conscious ones. The normal functioning of the body depends both on receipt of stimuli and on production of integrated responses. For the body to perform its activities in coordinated fashion, there must exist a connecting link between stimulus and response, between receptor organ and effector organ; that link must be a system capable of channeling information from one to the other.

As reasoning individuals we understand why we need a nervous system, but do we reason enough to understand how the nervous system works? The human nervous system is divided into the peripheral nervous system (PNS) and the central nervous system (CNS).

The Peripheral Nervous System: The PNS comprises all of the neurons and nerve fibers outside the CNS and has two subdivisions: the somantic nervous system, with nerves running directly from spinal cord to effector organs (actuators), and the autonomic nervous system (ANS) with nerves running from spinal cord to effector synapses (ganglia in humans is like an intermediate, N2 level bus in building system automation) outside the CNS and these send signals through nerve fibers to the effector organs.

The CNS consists of the brain and the spinal cord. It serves as a clearinghouse for all nerve impulses, controlling, directing, and integrating all messages within the body. The brain is about the size of a cauliflower and has been described as a "great raveled knot," a "modest bowl of pinkish jelly," or a "messy substance." The CNS is concerned with integrating and coordinating all nervous functions, both voluntary and involuntary. All sensory input arises from the environment (external and internal), and the effector organs (muscles and glands) are the ultimate destination of nerve impulses.

Lying between the CNS and the environment and the CNS and the effector organs is a vast network of nerves and ganglia that constitute the PNS. Hardly any tissue or organ is missed by the complex array of nerves, and thus, through them, the CNS is in continuous contact with nearly every part of the body. But there are other nerve components whose functions ordinarily lie outside our consciousness or control. These components connect with our visceral organs (heart, lungs, kidneys, blood vessels, intestines), and we neither are aware of what these organs are doing nor have voluntary control over them.

Most of the nerves are mixed and so include both sensory and motor components. The motor components of these visceral nerves are extremely interesting and important. They constitute what is known as the *autonomic nervous system.* There are two divisions within the ANS—the sympathetic and the parasympathetic. See Figure 14.8 and compare with Figure 14.9.

The sympathetic system mobilizes resources of the body for an emergency, effects that are sometimes called flight-or fight responses. Thus, in excitation or stressful situations (fear, anger, and the like), the heart beats faster, the stomach muscles relax, the pupils widen, and energy is made available because the brain sympathizes with organs. The *parasympathetic nervous system* is an energy-conserving system; its nerves are involved in relaxation and maintenance activities such as slowing the heart rate, enhancing digestive action, causing blood sugar levels to drop, etc. The condition of a particular organ receiving neurons from the autonomic system is determined by the relative amount of stimulation from the parasympathetic and sympathetic nerves. As a consequence, in the heart, many glands, and smooth muscle, which are innervated by both sympathetic and parasympathetic nerves, a dynamic seesaw balance in organ function is attained.

This is all interesting and rather obvious. But what does it have to do with the building system technology that we look at and deal with every day? Well, let's start with something relatively simple. How does the "intelligent" building control think? Not exactly like humans.

Conventional digital technology is based on bivalence. This means two absolute opposites—yes or no, on or off, 1 or 0, are or are not. Human logic functions with multivalence, which means that things can be partially so, to some degree, and partially not so, to some degree. For example, we can think of temperature as ranging from cold, to cool, to comfortable, to warm, to hot without knowing exactly what the temperature is. All building environmental systems operate on the principle of measuring exact temperature and then making a decision to cool or not cool. To take another example from everyday activity, say you are driving and need to make a turn because the road ahead of you is curving. Does your mind analyze the circumstances and immediately advise you: the road is curving with a 22° radius, requiring 11° clockwise initial rotation of steering wheel for a duration of 22 seconds? This may be mathematical logic, but our mind does not function this way to make our bodily systems react. In fact, our mind simply sees the road curving and autonomically makes us turn. All of this happens and we are probably not aware of it. It was sort of fuzzy which is what new technology is trying to achieve and it's called fuzzy logic. Using human perception instead of two-state temperature control building environmental systems can provide a fuzzy-cozy feeling.

As another exercise in logic, consider an apple. A nice, shiny red apple is clearly an apple in every sense of the word. Now take a bite out of the apple. Is it still an apple? Certainly—at least to some degree. Continue to eat the apple. Is it still an apple? At what point is it no longer an apple? In the precise world, an apple is a totally completed thing, and if any element is missing, it ceases to be an apple. In the real world, a partially eaten apple maintains its membership in the community of apples, at least to some degree. Remember, we live in a part-load world where peak load is a special condition. For the sophisticated user or designer who is big on personal comfort and individual control, fuzzy philosophy offers an attractive alternative mode of thinking and analysis. It is an important tool for bridging the gap between the science and art of engineering.

Perhaps age is an even better example. When does a person become an adult? According to law, it is at age 18. If we view this in a precise bivalent way, adulthood assumes yes/no characteristics. But is this reality? Is a person a nonadult up to the very last second before midnight of his 18th birthday, and is he fully an adult in the flash of time after the stroke of midnight? We all know some individ-

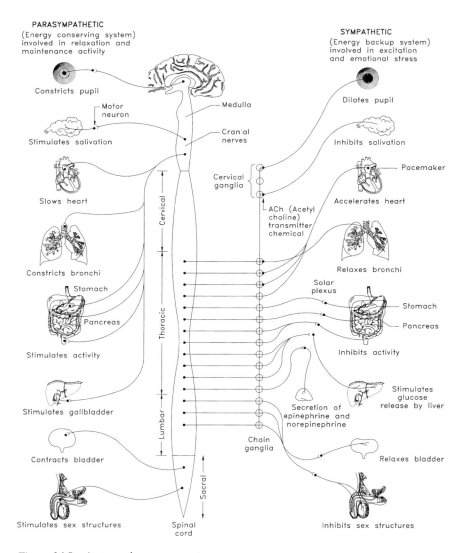

PARASYMPATHETIC
(Energy conserving system)
involved in relaxation and
maintenance activity

SYMPATHETIC
(Energy backup system)
involved in excitation
and emotional stress

Constricts pupil

Motor
neuron

Stimulates salivation

Medulla

Cranial
nerves

Slows heart

Constricts bronchi

Stomach

Pancreas

Stimulates activity

Stimulates gallbladder

Contracts bladder

Stimulates sex structures

Cervical
ganglia

ACh (Acetyl
choline)
transmitter
chemical

Cervical

Thoracic

Lumbar

Sacral

Spinal
cord

Solar
plexus

Secretion of
epinephrine and
norepinephrine

Chain
ganglia

Dilates pupil

Inhibits salivation

Pacemaker

Accelerates heart

Relaxes bronchi

Stomach

Pancreas

Inhibits activity

Stimulates
glucose
release by liver

Relaxes bladder

Inhibits sex structures

Figure 14.8   Autonomic nervous system.

uals mature more rapidly than others and that adulthood, even for the most intelligent of individuals, occurs as a result of a more gradual process.

While the technology associated with human logic is of growing and perhaps
prime interest to the building system engineering community, in many ways the
whole philosophy of fuzzy thinking is of far greater importance. Charles Ketter-

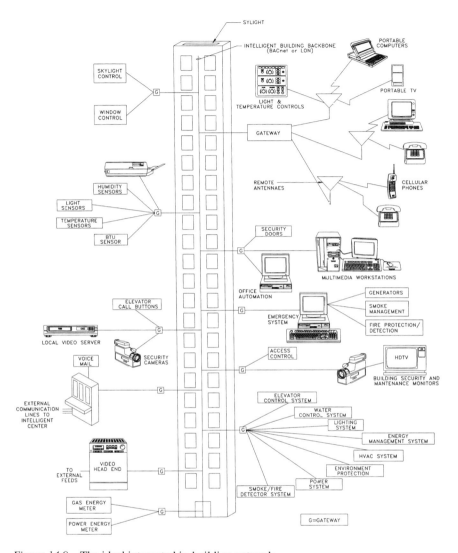

Figure 14.9 The ideal integrated in-building network.

ing once noted: "The only difference between a problem and a solution is that people understand the solution." We really don't understand the problem and as a result never get to a real solution. Applying fuzzy logic in this manner allows the designer to transcend building system science and to explore, and perhaps comprehend, the *art* of building systems engineering.

14.5.1   Human Logic versus Building Control System

How do our bodies integrate action? Two systems in our bodies act as coordinating links between stimulus and response: the endocrine system and the nervous system. The endocrine system regulates the activities of cells by means of hormones. The hormonal system provides for slow communication on a long-term basis. This is similar to time delay control action in building logic control. The nervous system, on the other hand, provides for rapid communication between the various tissues and organs of the body. The nervous system employs electrochemical messages, nerve impulses, that run along specialized nerve pathways receiving and transmitting information to and from various organs. Nerve cells, or neurons, are fundamental units of the nervous system, specialized to conduct electrochemical messages at high speed. Building systems communication protocols are based on neuron chips on the same guidelines popularly known as LON (Local Operating Network), a trademark of EcheLON. Neurons, when bundled together in cablelike form as nerves, are able to regulate the direction in which information flows.

There are three different functional classes of neurons: *Sensory neurons* receive stimuli from the environment and transmit information to the central nervous system (brain and spinal cord). These are comparable to transducers used in building automation systems. *Motor neurons* conduct messages from the brain and spinal cord to the glands and muscles, These are comparable to transmitters used in building systems. *Interneurons* act in an integrative capacity and shuttle signals back and forth between the neurons of various parts of the brain and spinal cord; this is where building systems technology, unlike body systems, is not coming together to share the information. In the body system a signal travels upward via the spinal cord interneurons and through a number of relay centers in the brain before reaching the higher centers (see Figures 14.10 and 14.11).

For example, let's analyze for a moment a motor that drives an air conditioner or perhaps more appropriately an air handling unit. For simplicity, let's assume that varying the motor speed varies the cooling output. Human logic will integrate this with the feeling of warm, cool, or hot. For example, from temperatures of 65 to 85 °F, we live in the category of warm, and say from 50% motor speed to 90%, our motor possesses membership in the category of fast. Thus, the human logic rule is *if warm, then fast.* So we have the beginning of a human logic system. Carry the analysis further, *if cold, then stop* motor and *if hot, then go maximum speed.* Notice how the control strategy has developed regions where a human rule applies and we can develop a family of rules that in effect give us a human logic system.

Human philosophy is also capable of assisting in the understanding of part-load conditions. Consider the initial step in an engineering analysis of a building system. Our prime objective is to determine the peak load and design a system around it. In conventional digital approach, we have two conditions, *peak* and *not peak.*

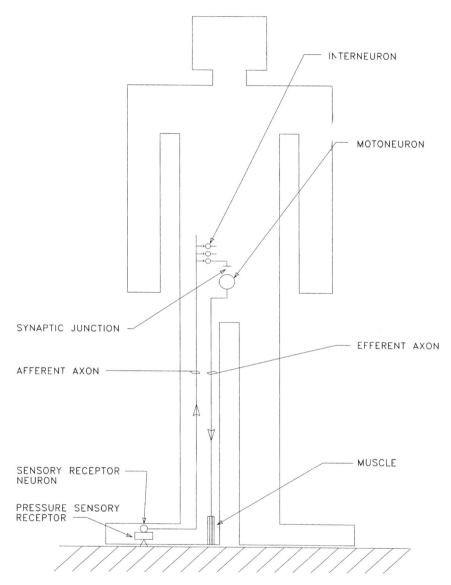

INTERNEURON

MOTONEURON

SYNAPTIC JUNCTION

AFFERENT AXON

EFFERENT AXON

MUSCLE

SENSORY RECEPTOR NEURON

PRESSURE SENSORY RECEPTOR

Figure 14.10 Excitable tissue is called into play when a person steps on a sharp pebble. Illustrated is a reflex arc that follows the excitation sequence: sensory receptor to its neuron to afferent axon to interneuron to motoneuron to efferent axon to muscle. The person starts to jump off the noxious stimulus in 0.025 second.

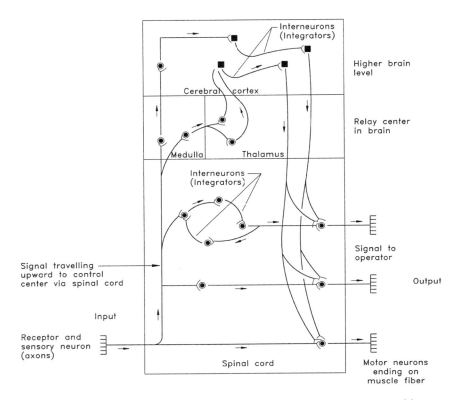

Figure 14.11 Simplified circuitry of the nervous system. Neurons are arranged into cables consisting of many axons (message receivers) and dendrites (passes signals). Axons are bundled together, to form a multistranded cable, from the nerve fiber or nerves. The collection of axons and dendrites in the brain and spinal cord comprises the information centers.

Suffice to say that peak conditions rarely occur. Applying human philosophy to the problem yields another approach, which recognizes that peak work load conditions are a special case similarly Zero work load is also a special situation.

Several observations about human logic (technically called fuzzy logic) are important. First, these systems depend on good input information, so sensing technology comparable to a body system is required. But rather than trying to model the control strategy mathematically with complicated equations, it uses a simpler series of fuzzy rules to develop a fuzzy control system. Since it is more human, it can be used by the less trained, and it is more easily maintainable. Building control systems based on fuzzy logic are especially helpful where nobody knows how to control the system, since all it requires to operate is basic intuition.

15

Integrated Building Systems Automation (IBSA)

15.1 Introduction

The primary subsystems in integrated building automation systems are HVAC, energy management, and lighting control. Inclusion of security, life safety (fire alarm, fire control and suppression, plus emergency aspects of vertical transportation), material handling, maintenance management, data/audio/video communications, and some aspects of office automation is the trend of future (see Figure 15.1). What the rapid advances in microelectronics and computer technology have done is to make possible and practical detailed, multipoint monitoring and control in real time, the result of which is a highly cost-efficient and environmentally appropriate facility.

The advantages of such an arrangement are so great that retrofitting of existing buildings, which is obviously more cost intensive than new construction, has become a major industry. The single largest issue in existing and new building automation today is open protocols. At the heart of the debate over interoperability lie two issues: distribution of control and expression of protocol. The first "D" in DDC can stand for two concepts: direct or distributed. Direct (the definition accepted in the industry today) signifies that system devices are monitored and controlled by digital electronics. "Distributed" control means that system devices interact among themselves to control a mechanical system, without reference to an overseeing device. The more distributed a building control system, the more likely are issues of protocol to arise. In a centralized system, monitor and control points all answer to a single, central controller. This central organ is responsible for administration of all protocol layers. In a distributed system, a common protocol is made necessary because there is no overarching device to regulate the flow of information between points.

Figure 15.1   Integrated building system blocks.

The building control system industry uses various buzz words to advertise its products and nomenclature. Terms such as intelligent buildings, computerized building control, integrated computer control, supervisory data systems, integrated building control, facilities management system, and so on are used almost interchangeably. Not to stay behind and adding to the list, the expression *integrated building system automation* (IBSA) will be used. Although the most obvious control function is to maintain desired comfort conditions, provide safety, and extend availability of building systems, integrated controls also increase fuel/power economy, by promoting optimum interoperation, and act as a safety net, limiting or overriding building system equipment. They also eliminate human error and never fall asleep even during a power outage.

Individual controls can be classified as follows: *controllers,* which receive/ send, measure, analyze, and initiate action; *actuators,* which are the controller's working slaves, and in turn become the muscles of pieces of equipment; *sensors,* which are the controller's reporters and feed back status to help the controller to operate systems; *limit and safety controls,* which may function only infrequently, preventing damage to equipment or buildings; and *accessories,* a miscellaneous collection of connectors, timers, relays, displays, communication medias, etc.

Types of controls can be classified in various ways. A common distinction is by motivating or actuating force: electronic/electric; pneumatic (in which compressed air is the motivating force); hydraulic (in which some form of oil is the motivating force); and self-contained, including "passive" controls such as those motivated by thermal expansion of liquids or metals, pressure in the system, or flow in the system. Another way to classify control system is by the motion of the controller equipment; *two-position* or discrete systems are of the simple on–off type; *multiposition* systems have several varieties of "on" position, commonly used for separate operation of more than one machine; *floating* or analog controls can assume any position in the range between minimum and maximum; *central logic control* or energy management systems can be programmed to integrate the many aspects of building system into one or more decision-making units (integrated logic control allows for a variety of functions that not only optimize a building system's interoperability, but also can enliven its appearance); and *distributed control,* which replace one or more central controller, with many dispersed controllers requiring little, if anything, in the way of a distribution system. Today, individually controlled machines are far more common.

Total energy management is an attempt to recover a lower grade energy of waste heat and extract every iota of BTUs and interuse all energies available in the building systems. For total energy to be successful, there has to be a reasonably steady demand in the building for the power used or generated and also for the heat recovered. The exhaust heat recovery from the HVAC systems or engines that power the generators is in demand for either heating or cooling at any time of the year. Lighting and the demand for power by computers, electrical business machines, and the great multitude of electrically powered devices in modern buildings often make the call for power reliability and cooling a indispensable part of the system.

15.2   Integrated Automation System Arrangement

A block diagram of a general IBSA is given in Figure 15.2, and a schematic of individual systems and related devices is shown in Figures 15.3 through Figure

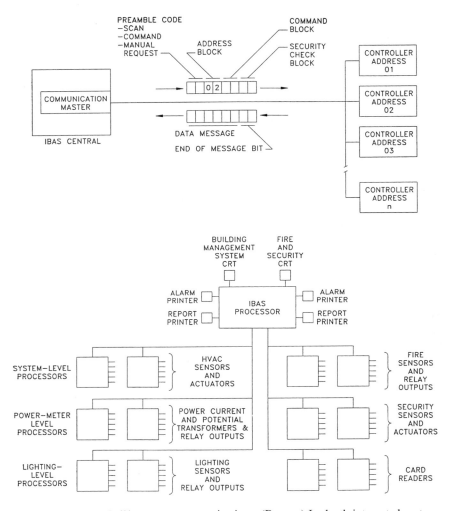

Figure 15.2 (Top) Poll/response communications. (Bottom) In-depth integrated system.

15.5. Any specific BAS may contain more or fewer monitors, sensors, dependent and standalone local controllers, and interconnected systems but the overall arrangement remains the same. Do not confuse this hierarchy scheme with protocol layers. Actually, depending on the equipment manufacturer and the system complexity, some of the functions can be integrated. Figure 15.6 shows development of integration associated controls and functions.

Sensors/transducers are at the bottom of the control chain. They contain automatic operation passive sensors such as flame, smoke, ionization, temperature,

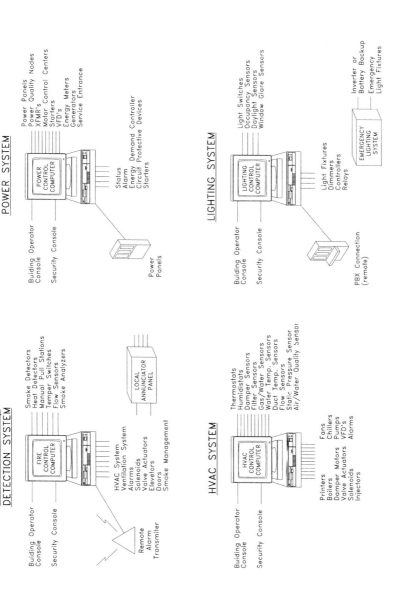

Figure 15.3 Intelligent building primary systems.

219

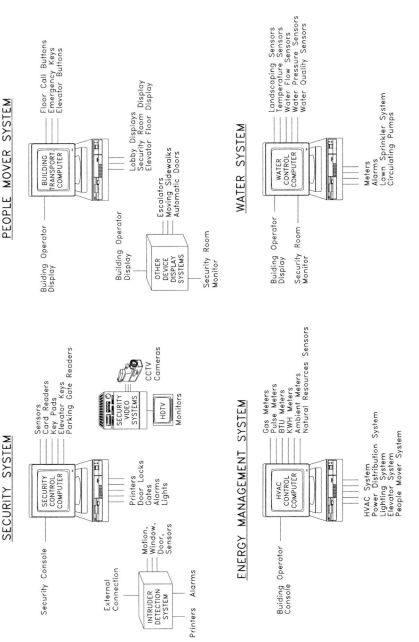

Figure 15.4 Intelligent building secondary systems.

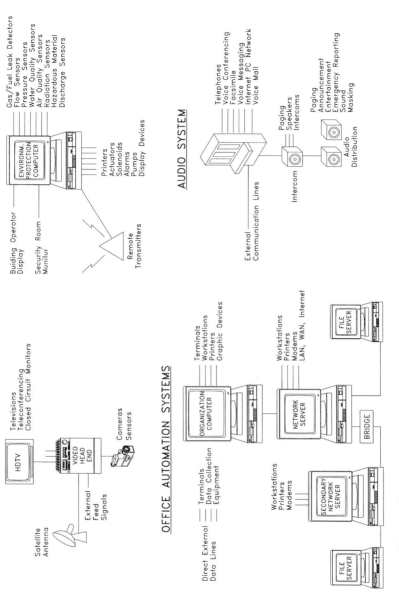

Figure 15.5 Intelligent building auxiliary systems.

221

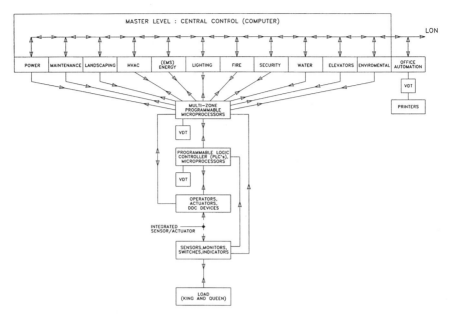

Figure 15.6 Integrated building automation system hierarchy.

lightning, vibration, daylighting, humidity, environmental, and water-flow detectors; physical condition indicators such as smoke and fire door position sensors; valve position switches, and manual fire alarm pull stations. All yield a digital signal (on–off) that is transmitted to controllers. The sensor/monitor level resides in load or with load to communicate to the controller the status of the temperamental load I call king and queen which the control hierarchy serves.

The level of sensor/operators or actuators (i.e., devices that both transmit a condition signal and receive an operation signal) could contain valve actuators, damper actuators, fire and smoke door actuators, fire system valves, power circuit breaker operators, and motor control (override) switches. This can either reside at this level or can be controlled through the building system via upstream levels. Control at this level is of a single device or block.

The controller level area is covered with either relay circuitry or microprocessor control (programmable controller), depending on system complexity. This level receives information from sensors and actuators, processes it, returns control information to operators/actuators, or operates specific local audible and visible alarms directly, and sends alarm and condition information to the building-wide system control and to the master controller (computer) at the top level. It also receives and processes signals from the central controller to ensure interoperation at an optimum level, alarm device operation for building emergency evacuation, and any reprogram or reset function.

The controller level contains some of the central level programming logic, or it can simply act as an input/output (I/O) device to a central computer. In a multi-building facility, this level contains the individual building system central control, interconnections to other building system controllers, as well as external connections for city alarm and external supervisory services.

Direct digital control (DDC) is more accurate owing to proportional, integral, and derivative loops and flexible than the proportional control commonly used in pneumatics. Such flexibility permits controllers to be tuned and allows control algorithms to be replaced or extended. For example, a DDC microprocessor can control temperature in a variable air volume terminal box, with control based only on a dry-bulb temperature sensor. The same microprocessor can be modified to monitor relative humidity and mean radiant temperature and to change the dry-bulb set point to better maintain comfort conditions. It can adjust airflows and temperatures based on occupancy indicators in an individual office. By transferring data back to a computer controlling the central fans, the microprocessor also facilitates minimization of the fan power required to deliver a given flow of air. Finally, it can control flow on the basis of air quality sensor measurement to provide a supply of outdoor air matched to the number of occupants in the space.

Microprocessor-based devices are currently used to turn on chillers and boilers at the optimal time for a building to recover from an unconditioned period. The required programs monitor the building's thermal behavior and adjust the equipment start time; such programs are examples of parameter estimation routines and give the controller an adaptive capability. This kind of optimization has been performed off-line and, more effectively, as part of an on-line control system. The control can be extended to include weather forecasting algorithms that can improve control of thermal storage or to schedule precooling by night ventilation.

Energy management devices are generally programmable controllers. Energy management is accomplished by controlling the functioning of most of the energy-consuming devices in the facility. These functions include duty cycling and load shedding and encompass HVAC, lighting, chiller control, and process equipment. Peak demand control and other energy functions may be included in this control device or may be accomplished by separate, dedicated devices, which may be connected to the central controller.

The central computer console receives status reports on individual devices from levels below it. Hard-copy printouts of all alarms and periodic status reports are generally made at this level, although they can also be accomplished at a level below. Video display terminals (VDT) at other levels can be used to view the status of any area and all the devices in the system, graphically or tabularly. Pictorial graphics for nontechnical end users are also possible. At this level interconnection is made to HVAC, elevator, security, energy management, maintenance, landscaping, power, environmental, and lighting systems in preprogrammed control and alarm modes, with manual override possible. Thus integration is achieved at this level.

All the signal systems that once were separate and distinct are now frequently combined and serve multiple purposes. For example, a fire detection might interconnect with the HVAC, elevator, security, water, electrical, and lighting systems to activate exhaust and pressurizing fans and deactivate supply fans, place elevators in fire evacuation mode, override access barriers to permit unhindered facility evacuation, connect fire pumps to emergency power feeders, disconnect high-voltage feeders, and activate emergency lighting, all via preprogrammed intersystem functions. Similarly a standby generator test can be conducted periodically via preprogrammed maintenance function. The system operator could then view any part of any system and reactivate or override programmed functions as desired.

15.3 Programmable Logic Controller (PLC)

The first programmable controller (PLC) design was introduced in 1969 as a replacement for the massive hard-wired relay panels then used in building system logic ladder diagrams to control system operations and components. The key feature of the new device was the fact that, as its name suggests, it was programmable. The additional features of modularity, expendability, diagnosis indication, and reliability in a building system environment also contributed to the PLC's success as an easy-to-maintain, adaptable control system.

Although originally designed for on–off applications, such as controlling the starting and stopping of motors, or actuator drives, PLCs rapidly spread to more sophisticated applications, such as those in the building system control. The evolution of the PLC over the past 25 years, owing primarily to advancements in microprocessors, high-speed communications, networks, and software, has gained the PLC a reputation as a reliable source of building system integration.

The platform of the PLC has two main sections: a central processing unit (CPU) and an input/output (I/O) interface section (Figure 15.7). The CPU is further divided into three components: the processor, memory system, and system power supply. The PLC also may include a peripheral programming device and an interface to a data communications network.

The processor provides the intelligence to command and govern the activities of the entire PLC system. The processor module contains the PLC's microprocessor, its supporting circuitry, and its memory system. The main function of the processor is to analyze data coming from all systems through input modules, make decisions based on the user's defined control program, and return signals back through output modules to the operating devices. The processor module

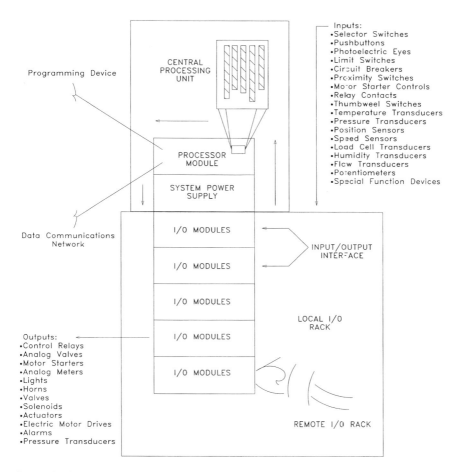

Figure 15.7 Integrated central processing unit.

also includes diagnostic indicators designed to detect communications failures as well as other failures during system operation.

The system power supply provides the voltages needed to run the primary PLC components (i.e., the processor module, memory system, I/O circuits, etc.). A battery backup, which provides power to the processor memory in case of a system power outage, is typically included in the PLC's processor module. The memory system in the processor module has two parts: a system memory and an application memory. Each of these two memory parts can be further broken down into individual memory areas.

The system memory includes an area called the executive, composed of a collection of permanently stored programs that direct all system activities, such as execution of the user's control program, communication with peripheral devices, and other system activities. The system memory also contains the routines that implement the PLC's instruction set, which is composed of specific control functions such as logic, sequencing, timing, counting, and arithmetic. A scratch pad memory area in the system memory is used to temporarily store a small amount of data for interim calculations or control. Except for this scratch pad, a system memory is generally built from read-only memory devices.

The application memory is divided into the data table area and user program area. The data table stores any data associated with the user's control program, such as system input and output status data, and any stored constants, variables, or present values. The data table is where data are monitored, manipulated, and changed for control purposes. The user program area is where the programmed instructions entered by the user are stored as an application control program.

The I/O interface section of a PLC connects it to external field and building floor devices. The main purpose of the I/O interface is to condition the various signals received from or sent to the external input and output devices. Input modules convert signals from discrete or analog input devices (such as those listed in Figure 15.7) to logic levels acceptable to the PLC's processor. The processor, based on decisions made by the user's control program, then returns corresponding output signals to the output modules, which connect them to levels capable of driving the connected discrete or analog output devices.

Input and output modules are housed in the same racks or panels that house the other components of the PLC system. The master rack, which contains the processor module, may or may not have room for all the I/O modules needed for the system. Additional I/O modules can be housed in a local I/O rack, which can be placed up to several hundred feet from the master rack, or in a remote I/O rack, which can be several thousand feet from the master rack. The remote I/O rack may also be used to communicate I/O information and the diagnostic status of remote field devices.

Every I/O module in a PLC system has its own address, and these addresses are used in the user's control program to identify each input or output device. The overall organization of the various memory areas within a PLC is referred to as its memory map. The memory map shows not only what is stored in the PLC's memory, but also where data are stored according to the PLC's specific I/O addresses.

At one time, most PLCs were programmed with dedicated loader/terminals. These devices have mostly been replaced by personal computers. A PC running one set of software can be used to program the PLC, and with other software can be used as an operator interface.

A data communications network may be connected to the PLC's processor module to allow communications to other control systems or computer networks. Data communications devices allow the central database (control program and system data) to be transferred from the PLC to a supervisory computer when required.

15.3.1   PLC Programming

Various languages are available to program PLCs, but for the past 25 years, the predominant language used to write PLC programs has been the relay ladder logic diagram. This type of diagram evolved directly from the wiring diagrams used for hard-wired replay panels before the introduction of the programmable controller. The use of familiar relay ladder symbols and expressions to define control logic for PLCs provided an easy transition for process operators and maintenance personnel who, most likely, were not trained in the more high-level computer languages. Standard symbols for objects such as switch contacts and relay coils were adapted so that they could be keyboarded on a loader/terminal unit.

For example, the notation shown in Figure 15.8a represents the same hard-wired ladder rung shown in Figure 15.8b. This line of ladder logic includes the same components, a normally open switch contact $(-][-)$ at the left, a normally closed contact $(-]/[-)$ in the middle, and a relay coil $(-()-)$ at the right. The numbers associated with the symbols are input/output addresses. When a number of ladder logic lines are combined in sequence to form a program, each contact and coil address can be used throughout the program whenever that particular relay condition needs to be evaluated.

As the capabilities of the PLC grew beyond the contacts, coils, timers, and counters of relay systems, special function blocks were added to the language. Function blocks are software objects representing specialized control functions.

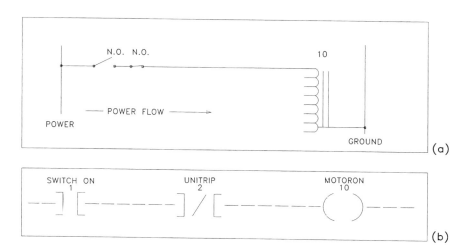

Figure 15.8   In this ladder wiring diagram the state of the switch contacts within the rung determine whether or not power will flow through the rung at any given moment. Switches 1 and 2 represent hard-wired connections to system inputs, while relay call 10 is connected directly to a system output.

A user can apply the same control functionality over and over again by encapsulating it in function block form, storing the function block in a library, then installing copies of the function block as many times as necessary in control programs. These special function blocks allow the user to program the PLC for various advanced applications, such as proportional–integral–derivative (PID) loop control.

With the addition of analog capabilities and function blocks, the boundaries between the PLC and process controller began to fade. The PLC's ability to combine binary operations (open–close, on–off) and analog functions (time, setpoint, position, level) give it a lot of flexibility in implementing building system control.

The PID functions of PLCs can be used to regulate temperature, pressure, flow rate, volume, velocity, or position in building control applications. Other PID capabilities of PLCs currently include bumpless source transfer, reverse and direct-acting modes of control, anti-windup protection, high and low alarm outputs, and the use of engineering units with a programmable scale.

15.3.2   PLC Communication

PLC communications include communications between: PLC modules within a rack; the master PLC rack and the local/remote I/O racks; the PLC and any connected peripheral equipment; and the PLC and a data communications network, which may be connected to a host computer, other PLCs, or other control devices.

As control systems become more integrated, they require more effective communications schemes between the various system components. Control systems require a number of PLCs to be interconnected so that data can be passed among them to accomplish the control task efficiently. Other systems require a communications scheme that allows centralized functions, such as data acquisition, system monitoring, maintenance diagnostic, and interoperating reporting, to be designed into a system to provide maximum control and interoperability.

15.4   From Building Technology to Microelectronic Technology

The single largest issue in building automation today is open protocols. At the heart of the debate over interoperability lie two issues: distribution of control and expression of protocol. The more distributed a control system, the more likely are issues of protocol to arise. In a centralized system, monitor and control points all answer to a single, central controller. This central organ is responsible for administration of all protocol layers. In a distributed system, a common protocol is made necessary because there is no overarching device to regulate the flow of information between points.

15.4.1   Protocols

When communication errors are a concern and/or multiple devices are using a network to exchange information, a protocol is needed to check the data for errors, to connect the data if errors are found, to provide equitable access for all devices on the network, and to ensure that only one device transmits at a time. A communication network limited to an office, a building, or several buildings close together is called a local area network (LAN). Workstations on a LAN can even share processing power. LANs vary in the type of medium (twisted pair, coaxial, fiber optic cabling or wireless), interconnection scheme or topology (bus, ring, or star configurations), and access method (contention, token passing, or time slice). An operating system can be added to the normal computer operating system, which handles the communication functions on the network. The topology of a LAN defines the geometry of the network, or how the individual nodes are connected to the network. The major factors affected by topology are data throughput (time taken between a change in state of a field input device and the resulting change in a corresponding field output device), implementation, cost, and reliability. The basic topologies used today include point-to-point, star, common bus, and master/slave (Figure 15.9). With twisted pair wiring and coaxial cables, information can be transported throughout a LAN at speeds of 10 million bits per second (Mbps). With fiber optic cabling, speeds can reach 100 Mbps.

Gateways are devices that interconnect similar or dissimilar LANs, dissimilar large computers, LANs and large computers, and so forth. A wide-area network (WAN) connects disparate elements, which can be LANs, over a large geographic area. A mainframe database serves as a repository for information available to numerous users. Each of these users may have different needs and require different interface options. Gateways to the mainframe repositories may be via modem, a network connection, or a direct connection. These gateways have numerous protocols depending on the class of computer being connected.

In the building systems environment the desire for interoperability drives the decision to tie systems together. Integrating systems requires many decisions. First, determine which of the many protocols existing today will be used to achieve interoperability. These protocols fall into three classes: proprietary, open, and standard.

Proprietary Protocols: To achieve interoperable communications with proprietary protocols, a translator or gateway must be created. Interfacing with proprietary protocols is typically an expensive, high-risk means to achieve system interoperability. However, it might be an expedient solution in certain applications.

Open Protocols: Open protocols share some characteristics of both proprietary and standard protocols. Interfacing to an open protocol system still requires a microprocessor-based translator. The communications methodology of these protocols can be easily supported by a cross-section of the industry as a standard part of products and does not require using a special translator or interface.

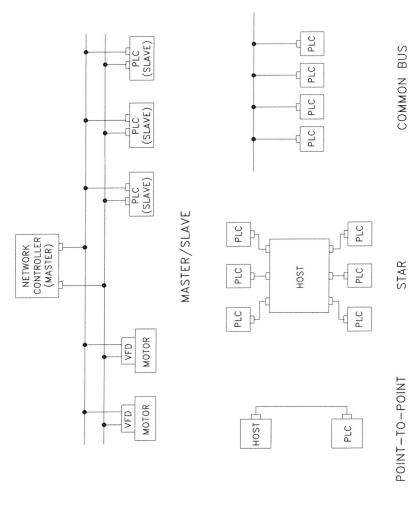

POINT—TO—POINT   STAR   COMMON BUS

MASTER/SLAVE

Figure 15.9 Local area networks used for PLC communication can have several topologies: point-to-point, star, common bus, and master/slave.

230

Today several standard protocols exist. One difficulty lies in the highly proprietary nature of all of this automation equipment protocol, making integration between systems and future alterations difficult if not problematic. Second, the resistance of some major manufacturers to open protocol (i.e., interequipment communication compatibility) places another difficult problem into the design process. BACnet, the building automation and control networking protocol in ASHRAE Standard 135–95, provides a means for building automation systems from various manufacturers to share data and interoperate. It is based on The International Standards Organization (ISO). ISO has defined the items necessary for a protocol in its Open Systems Interconnect (OSI) model (Figure 15.10) Layer 7 addresses application. Layer 7 is what we see when we use a computer program such as "Windows®" on a LAN or work with a PID control algorithm in a building automation system (BAS).

Layers 1 through 6 define the workings of a protocol. Here, data are interpreted, encoded, and transmitted over hardware. In a control network, much of the activity, and an estimated 90% of software development, is concentrated in the first six layers. You can see what a task it would be to embed the protocol layers—or even portions of the OSI model—in every point in a control system. Intelligence, traditionally, equals expense. Pushing that level of intelligence down to every point was traditionally considered cost prohibitive.

What traditional building control systems do instead is to confine network protocol issues to a few controllers. These controllers send data over a network. They then interpret received messages and communicate them to finite set of slave points below them see (Figure 15.11).

In the example in Figure 15.11, Controller "A" communicates with Controller "B" over a LAN. Let's say Controller "B" governs a chiller plant. It receives a signal that the building requires more chilled water. It would read that signal and then send a message to Point b, the drive on the chilled-water pump, ramping it up. Points a and b do not communicate directly. This system of control can be very effective in both performance and cost. A high-speed LAN such as Ethernet can carry large amounts of data very quickly. High-level controllers can be packed with great computational and data storage power. And making the choice to not build intelligence into a low level is a sound strategy.

Linking Controllers: When the industry talks about open protocol, it focuses on communication problems between these controllers. For example, let's say the building in Figure 15.11 adds a thermal storage system or power system or fire protection system with its own controller. The facility engineer wants to take advantage of the expertise that the thermal-storage equipment manufacturer has built into its control strategy. But the controller (C) uses a private protocol.

In order to meld the new thermal storage unit into an existing BAS, the facility engineer would have to coax or coerce the supplier of the thermal storage system and the supplier of the BAS to work together to write a software gateway be-

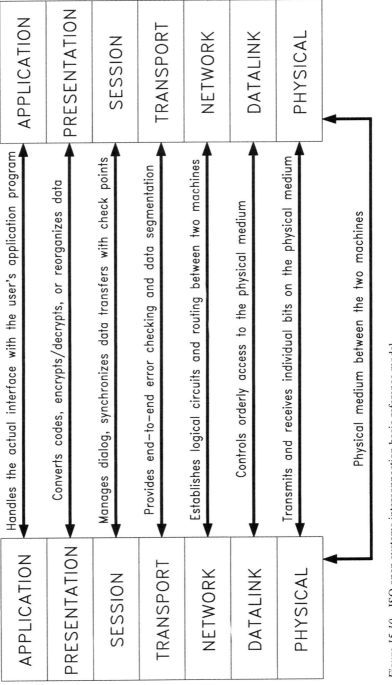

USER 2

APPLICATION
PRESENTATION
SESSION
TRANSPORT
NETWORK
DATALINK
PHYSICAL

Handles the actual interface with the user's application program

Converts codes, encrypts/decrypts, or reorganizes data

Manages dialog, synchronizes data transfers with check points

Provides end-to-end error checking and data segmentation

Establishes logical circuits and routing between two machines

Controls orderly access to the physical medium

Transmits and receives individual bits on the physical medium

Physical medium between the two machines

USER 1

APPLICATION
PRESENTATION
SESSION
TRANSPORT
NETWORK
DATALINK
PHYSICAL

Figure 15.10 ISO open systems interconnection basic reference model.

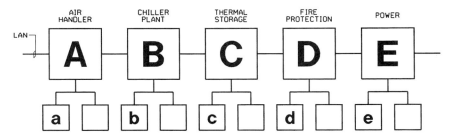

Figure 15.11   Traditional Local Area Network (LAN).

tween the two controllers. This gateway at LAN level would not change the control strategy in the thermal storage system, nor would it change the data gathered and interpreted at the BAS front end (both are Layer 7 functions). It would merely manage questions of protocol. A system programmer would still have to define the interaction between chiller, chilled-water system, and thermal storage system, on top of questions of protocol.

Common language protocol allows to use a single workstation for the daily operation of all building systems; and eliminate the need for redundant workstations from different systems. It integrates data from different systems and allows the user to graphically manipulate data for monitoring and control purposes. This is what BACnet is trying to achieve.

BACnet application layer protocol data units (APDUs) are divided into two classes: those sent by requesting BACnet-users (clients) and those sent by responding BACnet-users (servers). All BACnet devices are able to act as responding BACnet-users and therefore are prepared to receive APDUs sent by requesting BACnet-users. Many devices also act as requesting BACnet-users, and such devices are prepared to receive APDUs sent by responding BACnet-users. Both the requesting and the responding BACnet-users create and maintain a transaction state machine (TSM) for each transaction. The TSM is created when the transaction begins and is disposed of when the transaction ends. Vendors adopting BACnet as part of new systems are able to have interoperability without the need for any special devices. Interfaces to existing systems and non-HVAC systems might be accomplished today by using translators, similar to those described above. A BACnet translator provides a gateway to expand an existing system to many different BACnet compatible systems. When two BACnet compatible systems are connected, the owner, engineer, vendor, or system integrator has the responsibility to interpret the pieces in light of the specified functional requirements and determine a design for the system.

15.4.2   Integration of Network Objects

Objects are a logical representation of system data. The objects that form the basis of virtually all systems are Analog Input, Analog Output, Binary Input, and Binary Output objects. These represent the functionality of 0 to 20 psi, 4 to 20 milliamp or 0 to 5V inputs and outputs in a typical control system. Additional objects are Command Device; Event Enrollment; File; Group; Loop; Multistate Input/Output; Notification Class; Program; and Schedule. Each subsystem vendor should provide documentation as to the specific objects provided on the system, their function, and numerical ID.

Networking Options: Following networking options are supported by BACnet, depending on their performance requirements.

Ethernet (ISO-8802-3) LAN: Ethernet is a high-performance LAN with a data exchange rate of 10 million bits per second (Mbps). This network option is typically used for higher level system control in both office and industrial environments.

ARCNET (ANSI 878.1) LAN: ARCNET is also a high-performance LAN with a data exchange rate of 2.5 MBPS. ARCNET, which is a deterministic network, is widely used in industrial applications, and works well for building control applications.

BACnet Master Slave/Token Passing (MS/TP) LAN: MS/TP provides multidrop serial communications using an EIA-485 signal. MS/TP provides a low level of performance at a reasonable cost. MS/TP is typically used for communications to unit level controls.

Point-to-Point (EIA-232): Point-to-Point provides serial communications between two devices. It is typically used for dial up communications over phone modems. It may also be applied as a hard-wired interface. Point-to-Point provides a low level of performance at a low cost.

LonTalk: The LonTalk network uses a proprietary chip called a "neuron" for communications. LonTalk may be applied over a wide variety of physical media including twisted pair, power line carrier, and wireless. It provides a moderate data exchange rate at a reasonable cost, and is typically applied at the unit level. Devices using a neuron chip—sensors, actuators, and the like—can communicate to one another without reference to an overseeing controller (Figure 15.12). With the neuron chip, they possess a protocol, so they are able to send, receive, and interpret network-wide messages, instead of signals from a master controller. They have become "intelligent nodes" on a network. This network is what Echelon calls a LON.

A LON is easy to add to both because of the protocol and the way the network is programmed. Echelon's system is programmed in "Neuron C." The programming language is based on the ANSI C standard language, with additions to accommodate control functions. For example, it has a "when" statement to introduce temporal control and input and output data types. The lan-

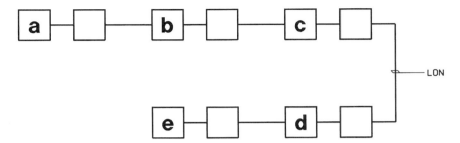

Figure 15.12 Local Operating Network (LON).

guage also has four different mechanisms for message passing between nodes. First, it can use an explicit message with a physical address. Imagine a supervisor telling a secretary, "Deliver this revised forecast to Mr. B." The second method of message passing is to send an explicit message with a logical address: "Deliver this revised forecast to office 2, 11th floor." The third is an explicit message with a named destination: "Deliver this revised forecast to the western regional sales manager." Each message-passing style broadens the parameters of delivery. The fourth method is to use an implicit message, called a network variable.

Let's take up the example of our building again, before the thermal storage system is added. Here, the sensor in the air handler is programmed to send an implicit network message, "need chilled water." That message is read by the drive on the chilled-water pump, and the request is answered by more chilled water. Now add in the thermal storage system. That same message, "need chilled water," could be read by the thermal storage system too. The thermal storage system would then react, as programmed. As in our example of a traditional network, a system programmer would have to define the interaction between chiller, chilled-water system, and thermal storage system. But in this system, no data gateway is necessary if all points on the system work with a neuron chip.

Table 15.1 lists key features of LON works and Foundation Fieldbus commonly used in operating networks in the industry.

A common network technology is a prerequisite before communications can occur between two systems. Special interface panels are used to convert between various physical links (LANs). Program and check out is required on each subsystem to achieve the optimum functionality. System integration in our example includes addition of fire protection, and power systems and this can be accomplished by one of the subsystem vendors or by an overall system integrator. They are responsible for reconciling device and network IDs and physical media. Once all of the subsystems have been properly configured and checked, the system integrator can connect the systems (see Figures 15.13 and 15.14).

Table 15.1   Networks—Comparative Analysis

Network characteristic	LON Works	Foundation Fieldbus
Technology developer	Echelon Corp.	Fieldbus Foundation
Year introduced	March 1991	1995
Governing standard	ASHRAE of BACnet	ISA SP50/IEC TC65
Openness	Public documentation on protocol	Chips/software from multiple vendors
Network topology	Bus, ring, loop, star	Multidrop with bus powered devices
Physical media	Twisted pair, fiber, power line	Twisted pair
Max devices (nodes)	32,000/domain	240/segment, 6500 segments
Max distance	2000 m @ 78 kbps	1900 m @ 31.25 kbps, 500 m @ 2.5 Mbps
Communication methods	Master/slave, peer-to-peer	Client/server, publisher/subscriber
Transmission properties	1.25 Mbps full duplex	31.25 kbps, 1 Mbps, 2.5 Mbps
Data transfer size	228 bytes	16.6 M objects/device
Arbitration method	Carrier sense multiple access	Deterministic centralized scheduler
Error checking	16-bit CRC	16-bit CRC
Diagnostics	Database of CRC errors and device errors	Remote diags, network monitors, parameter status
Cycle time: 256 discrete 16 nodes with 16 I/Os	20 ms	100 ms @ 31.25 kbps, <1 m @ 2.5 Mbps
Cycle time: 128 analog 16 nodes with 8 I/Os	120 ms	600 ms @ 31.25 kbps, <8 m @ 2.5 Mbps
Block transfer 140 ms of 128 bytes, 1 node	10 ms	36 ms @ 31.25 kbps, 0.45 ms @ 2.5 Mbps

15.4.3   Integrated Communication

Communication usually involves the transmittal of digital information. Traditional computer communications encompasses the connection of computer terminals to a time-sharing system, of word processing workstations to a central time-shared processor of one computer to another, and of a computer to various peripherals such as plotters, primers, and mass storage units.

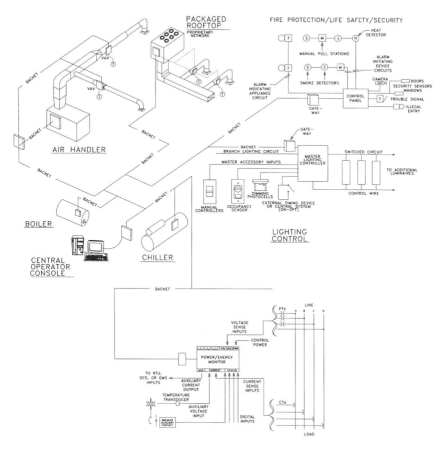

Figure 15.13   Single building systmes integration employing BACnet system.

In the building services industry, this definition is expanded to include connecting controllers to supervisory computers or other controllers, and connecting building systems to computers or fire or security alarm systems. Information passed includes sensor and status information needed for control or coordination, data logging, system tuning or maintenance, and set point or schedule information for performance modification or remote programming. Information is passed in bytes (groups of eight binary digits, or bits) which are the fundamental units of information.

The following are some of the benefits of integrated communications:

- Unit controllers may access a central time-sharing system when heavy computing is needed.

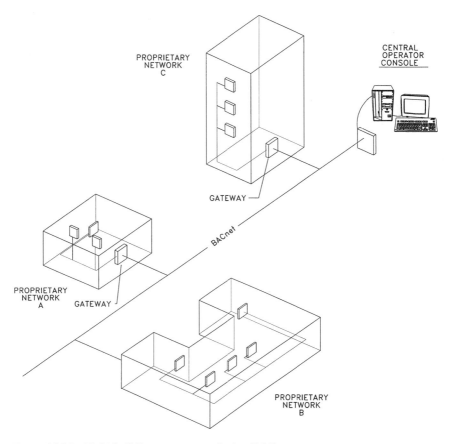

Figure 15.14 Multi-building campus employing BACnet system.

- Computers may be networked so that all computers on the network share storage and peripherals, as well as memory and computational capabilities.
- Building controls share occupancy performance, and set point information for control coordination energy-saving strategies, remote monitoring, and programming capabilities.

Generally, computers and peripherals communicate to each other through input/output cable connections called ports. A parallel port uses eight signal lines plus control lines for sending complete characters at one time. While this mode of transmission is quick, it is restricted to short (typically 15-ft) cables and cannot be used through telephone lines.

A serial port sends characters, one at a time, or serially, requiring as few as two or three wires. For serial communication, the receiver and transmitter must coordinate the beginning and end of each character, and the receiver must accommodate the bit rate (speed) at which the transmitter is sending information. To this end, communication can be achieved synchronously or asynchronously. Synchronous communications generally begin with a string of bits in a predetermined pattern. This pattern allows the receiver to synchronize with the transmitter. With asynchronous communication, each character has a specific length that includes a start bit and one or two stop bits. The receiver resynchronizes on every character sent. Although there is a time penalty of about 25% for sending start and stop bits for each character with asynchronous serial communication, most smaller computers do it because of the greater simplicity of both the hardware and the software required.

For communications over telephone lines, both asynchronous and synchronous systems can support dial-up connections. The voltage levels used for short-distance communications cannot be used with switched telephone lines or fiber optic cables. A modem (modulator/demodulator) interfaces the two media by converting bits between voltage levels and tones or light. The quality of the medium determines the communication speed. Current technology provides up to 38,400 bits per second (bps) over normal telephone lines, at 9600 bps with 4:1 data compression. Speeds of up to 1 Mbps can be supported over dedicated T1 telephone lines. As more information is passed around buildings, voice, computer data, building services data, and CATV networks may be called on to integrate in order to reduce installation and reconfiguration costs.

Linking local microprocessor controllers with a central, supervisory computer makes it possible to integrate environmental controls with such services as security, life safety monitoring, and lighting. The network can be extended to the electric utility, which provides spot pricing information as input to load-shedding programs, and even to dropping specified equipment in the event of a power shortage.

Central computers are already used to collect data from individual controllers or meters, but typically perform a minimal amount of data analysis. Analysis performed off-site by building service contractors identifies long-term trends in energy consumption, isolates beneficial or harmful changes in equipment operation, and normalizes energy consumption for changes in weather. These analysis programs can be incorporated into on-site computers, providing operators and management with up-to-date, readily available information.

Recently, use of the public domain, worldwide Internet has increased dramatically for such functions as e-mail and file transfer. The Internet was originally developed to allow information exchange between government and academic researchers. Does cyberspace globalize building systems? The word "cyberspace" can be best defined as "a consensual hallucination experienced daily by billions of building operators" or . . . . a graphic representation of data abstracted from every computer in the building system. As yet there is no universally agreed upon definition of cyberspace. We do know its basic building blocks are computers electronically linked into networks through which data can be exchanged at lightning fast speed. Cyberspace is where this occurs. More useful than a definition, however, is the significance of cyberspace to the System Engineering profession.

You need no longer go somewhere or have someone come to you to obtain or exchange information. For example, you can employ online historical databases or library operation/maintenance manuals. E-mail or groupware programs may substitute for face-to-face meetings. Reading World Wide Web home pages may substitute for visiting whatever entity has set up the home page. Although it may seem an exaggeration, it is possible to suggest that we can now "wander the fine architecture and systems of buildings on the earth and not leave home."

A virtual visit will not be the same as a physical visit and an online database is not the same as field data, but electronic connections may be a satisfactory substitute depending on one's needs. It is, of course, true that we have always had options for obtaining information at a distance. Rumor, gossip, telephones, and "word of mouth" have always been with us. Computer networks, however, make immediately accessible and easily usable information stored in connected computers and in the minds of building engineers linked to these computers. For those who have already experienced cyberspace, it is clear that the ability to interact at a distance via electronic means with building operators and informational sources is quite novel.

The most common attitude of computer users is that the computer is simply a tool; cyberspace, just one more electronic tool. If cyberspace is more than a new tool, however, it will affect not only efficiency but also the environment in which building operation is conducted. It will determine who has access to system information and how this information is valuable. It will not only provide us with information quickly but also change the rate at which knowledge becomes obsolete. It will, in other words, not be simply a useful technology but a defining technology that multiplied the wisdom of integration.

16

From Microelectronic Technology
to Information Technology

16.1 Intelligent Buildings

Although the term "intelligent building" seems to attribute a human characteristic to an inanimate object, the term's actual meaning, at least as defined by the Intelligent Buildings Institute, is "a building which provides a productive and cost-effective environment through optimization of its four basic elements—envelope, systems, services and management—and the interrelationships between them . . . Optimal building intelligence is the synchronizing of solutions to occupant and business need." In the framework of such a definition, the intelligence of a building resides essentially in its systems design. An intelligent building is, therefore, not necessarily packed with electronic systems, nor even one with an extensive Building Automation System, unless it is necessitated by the building program and can demonstrate its cost-effectiveness. Rather, it is a building designed with thought toward integration and perhaps some inspiration, both of which are necessary to satisfy immediate and predictable need, and yet also to anticipate the occupants' future requirements. Since modern building usage is computer oriented and rarely static, it can now be said that "It is possible to turn off the computer but, it is not possible to turn off the computer environment." Refer back to Figures 15.3, 15.4, and 15.5 to visualize how computers are all around us. Figure 16.1 shows how intelligence shifts in a mechanical workhorse just as in humans, and how a motor junction box carries not only wire terminals but also "brains" to control motor speed.

Experience has shown clearly, however, that in the very fast-moving building systems market, today's high-tech can become tomorrow's albatross, and therefore planning for future requirements means intelligent preparation as much as immediate provision. The integrated building is essentially the building that is designed with foresight. Hindsight will determine the degree of intelligence.

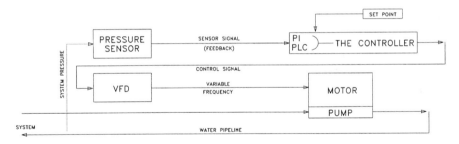

VFD SYSTEM—ALL COMPONENTS WIRED SEPARATELY

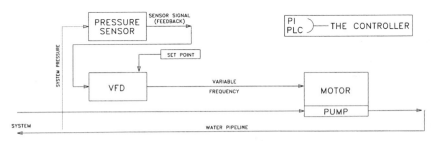

VFD SYSTEM—CONTROLLER INTEGRATED INTO VFD

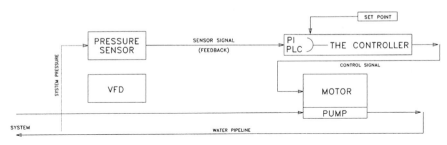

VFD INTEGRATED INTO THE MOTOR—CONTROLLER IS A SEPARATE UNIT

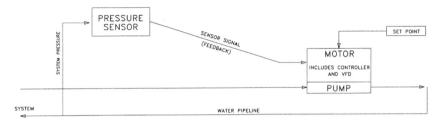

VFD AND CONTROLLER INTEGRATED INTO THE MOTOR

Figure 16.1   Integrated variable frequency drive motors.

16.2   Intelligence Communication and Dissonance

From smoke signals and Samuel F. B. Morse to Alexander Graham Bell and Guglielmo Marconi, to talking around the world and to the moon, the history of communications has been one of overcoming distance. The decibel unit is of special interest because it is so commonly used in communication work and because it has several applications. The first application is from the original definition of the Bel, named after Alexander Graham Bell. The Bel is the log (base 10) of a power ratio. This number is generally too small, so the decibel came into being, namely, 10 times the log (base 10) of a power ratio.

$$dB = 10 \log_{10}(P_o/P_i)$$

It is generally used to define the gain or loss of a device. If P_o is the power *out* and P_i is the power *in,* then the ratio P_o/P_i is the response of the system. The log of the response allows very large or very small responses to be easily written.

A common error is to read a large system gain number in a specification without the input specified, and then to assume that the output power will also be large. However, system inputs are usually limited to a fixed maximum, so the output may not be as large as suspected or desired.

Example: If $P_o = 10$ watts and $P_i = 1$ watt, the system gain in dB is

$$dB = 10 \log(10/1) = 10 \text{ dB gain}$$

However, if $P_o = 100$ watts and $P_i = 10$ watts, the system gain is

$$dB = 10 \log(100/10) = 10 \text{ dB gain}$$

Note: the system gains are the same with a 10 times difference in output power.

Today technology has minimized the practical distance limitation, although for specific communication requirements distance is still the major key to finding the proper system. The large number of choices alone makes it very difficult for the noncommunication type building system engineer to choose confidently. All intelligence communication systems can be divided into five basic pieces. The information source and the information user mark the beginning and the end. Between these are the two signal conditioners and the medium that separates the source from the user.

Information sources of building systems are typically pressure, temperature, current, flow, voltage, voice, and data. Typical information users are charts, computers, printers, video display, alarms, and loudspeakers. The signal-conditioning requirements and the transmission medium determine the communication system itself. Common conditioning devices are modulators, analog-to-digital or digital-to-analog converters, modems, and isolators. The medium may be a simple wire (twisted/untwisted), coaxial cable, free-space, a radio system, a satellite, or a fiber optic cable.

An electrical signal is often the reasonable choice to transfer information when the distance or obstacles preclude directly coupling the information source and the information user. An example of a directly coupled system is two people talking, the human voice (the source) acoustically coupled to the human ear (the user).

Dissonance in intelligence is caused by electromagnetic interference and covers the frequency spectrum from dc to the optical frequencies (30,000 Ghz). In recent years, many buildings have been upgraded to intelligent buildings, resulting in controls upgrade from analog type to digital type. Yet, many systems have both analog and digital components. The low-level, narrow-band analog has given way to the high-level, high-noise immunity logic families. But the bandwidth of the logic is orders of magnitude wider than that of the analog circuits, and therefore more sensitive to transients and Electrostatic Discharge (ESD). The Electromagnetic interference (EMI) problem continues because of the system grounding, large number of ground loops, and long cable runs.

The most frequent EMI violations in grounding of control systems today are: both commercial power and UPS safety grounds entering the cabinet, cabinet grounded to building ground grid and/or steel and indirectly grounded to other ground points, Internal dedicated analog reference (returns) grounded via long external runs to building ground, analog and logic returns connected together (often indirectly) and grounded to the cabinet, uncovered cable trays multipoint grounded to building steel, cable shields that are left unterminated at the point where they enter the cabinet, unsuppressed and unfiltered power mains at the cabinet entry point, and analog sensor and control device cables located in the same cable tray (see Figure 16.2).

If analog sensor and control device cables must share the same cable tray (a poor practice), at least position the sensor cables on one side of the tray and the controlled-device cables on the other side. Use tray separators. Since most cable trays, unfortunately, are open at the top, ferrites can be added every 10 or 20 feet.

16.3   Intelligence Source Objects, Quality, and Transmission

The device required to transform the primary source—pressure, flow, temperature, rpm, vibration, whatever—into an electrical signal is called a transducer. The definition of a transducer given by the Instrument Society of America (ISA) is a "a device which provides a usable output in response to a specified measurand." For our purpose the output will be the electrical signal that carries the desired information.

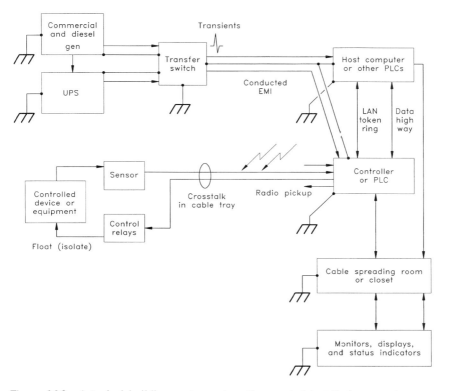

Figure 16.2 A typical building system automation control installation contains many ground loops and other ways for EMI to couple into equipment.

The transducer will almost always require an external power source, even if the sensing portion is "self-generating," for example, a thermocouple. This power requirement is often overlooked if the systems are not integrated and later requires a system "add-on." In addition, this power source must be considered in any reliability calculations made for the system. Three important characteristics to consider about the transducer are the output signal's type, power, and range. Equally important is how they interface with the next stage, that is, signal-conditioning equipment.

Intelligence Quality: The object of the signal-conditioning equipment is to prepare the signal to overcome the medium loss, overpower any undesired interference, and arrive at the user input terminals with less than or equal to acceptable distortion. Many times the conditioning done on the intelligent source end must be undone at the user end. Thus, the pairing is often seen in manufacturers' literature. Examples of these pairs are: (Digital to Analog-Analog to Digital) D/A–A/D convertors, modulators–demodulators, and rectifiers–inverters. A good example of this type of system would be a computer that is used to monitor the pressure of a pipeline. The pressure transducer is the source and the computer is the user.

Intelligence Transmission: By definition the transmitting medium will include everything between the two signal conditioners. This can be any system from a single wire to a satellite circuit. Therefore, the term "intelligence loss" can be misleading in that there could be signal gain devices included in the medium, so the user receive signal level could be higher than the source signal output level. Whether the medium presents a gain or a loss to the overall system, it is a piece of the problem that can be accounted for easily. The hidden problem, and one that is much harder to account for, is the medium's susceptibility to interference. The interference can be noise in any or all of its forms or a variety of unwanted signals. For simplicity, noise will be considered as any signal interference, including the desired signal interfering with itself as in the case of an echo. A large intelligence loss that requires a large signal output power is no different than a low medium loss with interference that requires a large signal output to override the interference.

Some building systems are more susceptible to a particular noise or distortion and, of course, the longer the distance between system components and controllers, the greater the exposure to these conditions. Because of the importance of noise considerations in system design, many volumes have been written and will continue to be written describing noise. Acoustically, common descriptions are hum, pop, rumble, click, etc.—typically, those things that are heard that interfere with the communication process. Visually noise is not so common, yet it is present. Glare and spots are considered visual noise to the eye and on a video display ghosts, snow, and tearing are caused by noise.

All mediums will distort the signal to some degree. Two things can be predicted about noise. The first is that it will always be present and the second is that its waveform is unpredictable. If the noise waveform could be predicted, then a 180° out-of-phase duplicate waveform could be generated and added, thus achieving a noise-free system. This is the ultimate ideal, in the same category as perpetual motion.

Noise, although difficult to describe, can be measured. The measurements vary as do the instruments used but generally root mean square (rms), average, rectified average, or peak voltage determines the quantity of noise present. These measurements depend on the filtering and weighting applied and, of course, represent only the amount of noise present at the time of measurement. The most important design criterion for most building intelligence communication systems is the signal-to-noise (S/N) ratio received at the user's terminal. Often it is used as a measure of quality of an overall system. In data systems bit error rate (ber) is the normal design criteria. In reality ber is also dependent on S/N.

Noise in intelligent buildings can be divided into two major groupings. The first is noise generated internally to the building system by design problems, integration difficulties, or a system operating at parameter thresholds. In any system design or device specification there are tradeoffs between the technical ideal and costs. The second category is noise encountered from independent external sources.

16.4   Intelligence Availability

In intelligence communication systems, failure should also include external source interference as well as actual equipment failure if the interference is sufficient to make the information transmitted unacceptable. Thus intelligent buildings communication can be lost as a result of (1) equipment failure, (2) loss of usable signal, or (3) loss of power. Although related, each of the three can to some extent be considered separately. Individual unit reliability or power reliability is covered under a separate chapter; therefore, further discussion is not necessary here. Outside influences such as corrosion, propagation failure, or noise problems, as a special case, are the focus here.

Corrosion protection, like lightning, is a study in itself. One of the causes is the galvanic corrosion occurring between two dissimilar metals touching each other in the presence of an electrolyte. The source of the electrolyte could be the humid atmosphere, soil, or the metal surfaces themselves. A lead-shielded communication cable buried in an electrolyte of soil and water carrying a current of 1 amp from the lead to ground can lose as much as 75 pounds of lead a year as corrosion to the ground.

A common problem in intelligence communication circuits is the aluminum junction box terminal strips connected to copper wires. Sometimes aluminum is used as grounding wire connected to a dissimilar copper-jacketed steel grounding rod. The typical method of protection is to maintain the junction as free from any electrolyte as possible. Coat the junction with corrosion-resistant material: paint, organic coating, or electrodeposits. This coating must be free of pores and discontinuities.

Corrosion can also be caused by stray dc currents passing through metal to ground. These currents can be coming from local power systems, other cathodic protection, etc. A common method for protection is to apply an external dc voltage of the opposite polarity to cancel the stray current. The buried metal is forced to be a cathode (negative relative to ground) instead of an anode. And the other metallic ground connection would be the sacrificial anode.

When currents are caused by fluids flowing through a pipe or other ground potential differences that causes a current flow from the metal to ground, corrosion takes place. The protection against this action is called impressed current. A voltage is applied to the system so the impressed current opposes the natural current. The applied voltage will cause a net current to flow from a sacrificial anode to the metal system, thus adding metal to the system instead of metal corroding away.

In intelligent buildings, the basic principles of external interference protection are:

Distance—can be controlled by construction and installation techniques

Barrier—can be provided by insulation/insulators, sectioning, shielding, metallic raceway

Training—to ensure maintenance personnel are familiar with the system so they are aware of potential failure to the operation

Interference—exposure protection often is the least expensive and may in some cases be mandatory. The National Electrical Code deals extensively with spacing requirements between communication circuits and power lines. Usually there are governmental regulations that determine minimum clearance for cables and wire that cross.

Maintenance: The building system engineer putting together an intelligence communication system to meet some need that is a small part of an overall building systems will almost invariably forget maintenance—until the system falls in some area. Maintenance plays a large role. It is discussed in detail in the reliability section. Maintenance will often be the controlling factor in making an economic choice among different types of systems.

Three major approaches are taken regarding systems maintenance in today's industrial world. Leasing versus ownership is one approach. If the system can be leased, then the routine and emergency maintenance will come as part of the system lease costs. A time factor may have to be considered by both parties; that is, the mean time to repair (MTTR) should be agreed to in the leasing contract. A lease versus ownership study would have to be tailored to each system requirement and is outside the scope of this book.

Another approach is to purchase the intelligence system units and, after the system is tested and commissioned, utilize a maintenance agreement with either the vendor or a service organization. The terms of such a contract are normally discussed in general with potential suppliers before the final selection is made. Along with the list of equipment is a description of the system, usually with a block diagram showing the overall intelligence signal flow.

16.5 Optical Fibers and Intelligent Buildings

Optical fibers have been used for years for guiding light in noncommunications applications such as medical illumination and image transmission. Introduced to the building systems communication industry in the mid-1980s, fiber optics received a tentative reception. That now has given way to the point that fiber cables are the medium of choice for many new and retrofit intelligent buildings information exchange or communication systems.

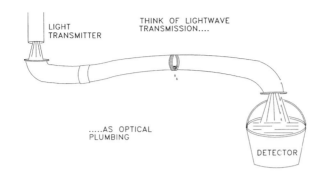

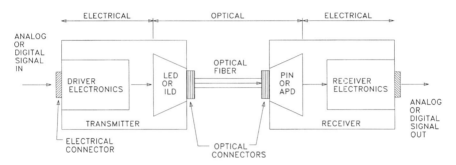

Figure 16.3 (Top) Example of lightwave transmission leakage. (Bottom) Basic light-wave system.

The term "fiber optics" refers to the technology of transporting information using guided light waves in an optical fiber (See Figure 16.3). The fiber's capacity for total internal reflection enables a beam of light from an optical source to travel very long distances and around bends. The transmission mechanism of fiber optics is simple. Light flows only in the center portion, or core, of multilayer glass or plastic fibers, which are contained by an outer layer or cladding. The flow rate (or power throughput) depends on the source power, input-coupling efficiency, size and physical condition of the fiber, leakage at connectors and splices, constrictions at bends, and output-coupling efficiency.

Optical-transmission system components include transmitters, receivers, and cables. A laser diode (LD), known as a coherent transmitter, provides a high degree of spectral purity and high power densities. LD signals exhibit nearly perfect beam collimation, meaning there is virtually no spread to the light beam. The alternative is a light emitter, which can be used for short transmission lines.

LDs and LEDs are currently the only practical light sources for fiber optic communication systems. Generally, LDs perform better due to their more tightly focused spectrum, efficient coupling into fibers, and high modulation speeds. On the other hand, LEDs are less expensive and simpler to operate, are less temperature-sensitive, and last longer. Whichever light source is specified, transmitters should produce more than a few milliwatts of monochromatic power efficiently and operate at or near room temperature with no special cooling. Long life, compactness, and handling convenience are also important considerations.

Receivers incorporate photodetectors, which convert optical radiation into electrical signals, and square-law detectors that respond to light intensity averaged over time. Avalanche photodiodes, one type of photodetector, have internal gain resulting in high sensitivity but also require a relatively large power supply. The alternative, positive-intrinsic-negative diodes, are less sensitive but may be a better choice when high sensitivity is not required.

Cables consist of optical fiber cores covered with cladding and sheathed in protective jackets. Fibers are available in various combinations of glass and plastic. Both core and cladding may be plastic, yielding a fiber with high losses but one that is easy to handle and economical for short links. Although plastic fibers do not have the dimensional stability and environmental durability of glass, they are acceptable for some applications. A glass core also may be clad in plastic for special applications, such as communication in high-radiation environments. All-glass fiber features low signal attenuation and probably is the best choice for long transmission paths.

Optical fibers may be either single- or multimode. Only one light mode can propagate in single-mode cables and there is no modal dispersion. These fibers, which have core diameters of 5 to 10 micrometers, also require single-mode light sources, that is, lasers, and are relatively expensive and difficult to couple. Multimode fibers allow more than one light mode to propagate and, with core diameters typically 50 micrometers or more, are easier to work with.

Splicing fiber optic cables provides reliable permanent connections with low signal loss but requires reasonable skill and special tools. Couplers are available to join or split optical signals. Plastic fiber ends can be cut with a razor blade, but glass must be polished to be effective. However, because glass accepts a better polish than plastic and shows little or no deterioration with time or wear, it usually is specified for applications requiring high reliability. As in wire-based systems, minimizing the number of connectors and splices reduces signal loss.

Multilane: Fiber optic technology can be applied to bidirectional and multidirectional communications networks. Two fibers, carrying signals in opposite directions, are used in the most straightforward scheme—referred to as half-duplex—for transmitting and receiving at both ends of a point-to-point link. A full-duplex system, permitting simultaneous transmission in both directions on the same fiber, conserves fiber, offering a significant advantage on long links. By applying special equipment, large amounts of data also can be multiplexed, or combined for transmission and separated at receipt, on fiber optic systems.

The advantages of using fiber optic cables over conventional copper cable, coaxial cable, or twisted-paid systems for intelligence communications are numerous:

- Optical systems do not generate electromagnetic interference and are not susceptible to outside interference.
- Complete electrical ground isolation is achieved between transmitter and receiver on an optical system, eliminating ground loops and common ground shifts in data circuits and permitting safe operation in hazardous environments while providing high-voltage isolation.
- Data security is increased because fiber optic transmissions do not radiate electronically detectable signals and are thus difficult to intercept. Building systems integrated with building security systems require this attribute.
- Optical transmission cable and equipment are generally smaller and lighter than what is required for comparable electronic systems. Furthermore, a single line can provide multiple control functions and carry audio/video signals.
- Cross-talk is eliminated on optical multichannel systems.

Disadvantages of fiber optic systems, while few, nevertheless should be considered:

- Not all building system equipment enclosures are able to accept fiber optic transmitters. A practical solution is to wire input/output points within a given perimeter zone with copper to a junction box near the zone equipped with a fiber optic multiplexer. The multiplexer accepts individual inputs and transmits them over a single fiber optic line.
- Each fiber optic transmission link requires a transmitter and receiver with corresponding power supplies, which means power must be available locally at all devices using fiber optic communications.
- Fiber optic splices require trained technicians and more sophisticated equipment than do metallic-conductor systems. Fiber preparation for outdoor terminations and splices can be difficult and may require a protective enclosure around the work area.
- Fiber optic equipment, including cables and connectors, is not as readily available as the components for metallic-conductor systems. This, however, is changing rapidly as equipment suppliers are beginning to stock high-quality components with more variety to ensure off-the-shelf availability.

17

Reliability Requirements, Risk Management, and Associated Building Systems Engineering

17.1 Introduction

A commonly used definition for reliability is: the ability of a system to perform a required function under known conditions for a desired period of time or a determined amount of usage. Closely related to reliability, and actually what the system engineer is interested in, is availability. Again, the definition is as follows: the probability that a system subject to repair will perform a required function under known conditions on demand. Notice that availability has introduced two additional concepts: (1) the probability of something happening and (2) repair.

In overall system reliability don't overlook the operating time. In building systems typically the standby generator operates for less time than the chillers. Two terms are useful: MTBF (mean time between failure) and MTTR (mean time to repair).

Quantitative reliability or availability figures are normally used for comparison purposes. Quantitative reliability consists of two major elements: (1) the probability of failure, and (2) equally important, the magnitude of the consequence of failure. If the system can fail with little consequence, then the quantitative reliability is different than if some failure resulted in major undesirable consequences. An individual building system equipment manufacturer is concerned about balancing the different component failure rates, so that the unit doesn't result in an obvious weak link. The same idea is true for the system designer; the system is no more reliable than the least reliable unit. The good design results in the least reliable link having the least consequence.

Common system protection is normally employed for three reasons: (1) personnel safety; (2) equipment safety; and (3) equipment reliability. Of course, what the designer is protecting against often determines what the protection is for. As an example, protection against accidental contact with high voltages is for personnel and equipment. The major category to be protected will determine the type of protection equipment. Personnel protection will be different from equipment protection even though they both may be protecting against high-voltage exposure. It becomes apparent that the designer must first decide what must be protected and what it must be protected from, in order to decide the type of scheme to implement. There are, of course, types of equipment to protect multiple categories as well as multiple causes. For many years when engineers talked about the reliability of building systems, they often used nonspecific quantifiers such as "good, better, best," or "what we propose is better than what you have," or "if you don't do something soon, it will probably fail and cause tremendous problems." If they did try and get more scientific, they usually would drag out the IEEE gold book or military handbook and dig through it, and then try to make heads or tails about how the reliability calculations worked, and then try to figure out which piece of information from the mounds of data in the book should be used to represent the problem at hand.

One of the key points emphasized here is the need to speak the language of the business community. In today's world, this language is centered around such factors as business risk, cost impact, first cost, life-cycle cost, and the cost to operate and maintain. Notice that the language almost always deals with cost and is based on actual data. It may well be a waste of time if reliability is discussed in a language other than cost and in a context based on something other than fact-based data. To have meaningful results from a system reliability evaluation, the answer must be in dollars. Develop a modeling tool that can be applied to both new and existing real life systems that will help determine the reliability at any point in the system and will determine the cost impact from loss of function. Figure 17.1 shows a reliability evaluation process that speaks engineering and business management language.

For example, if a building system model is analyzed to achieve 97% availability and it did not include planned turnarounds, which were estimated to be 2%, this will result in a 99% actual availability. Of the remaining 1%, 0.5% can be allocated to problems associated with human operating issues This leaves 0.5%, half of which (0.25%) can be allocated to management related problems and 0.25% can be allocated to building system problems. Using simple mathematics, 0.25% of one year is about 22 hours (say one day). To meet business requirements, the impact of all building system failures to the operation of building functions had to be less than one day per year. "What if" analysis could be performed to see the impact of component on system structure reliability.

17.2 Basic Reliability Terminology

Annual Risk: The calculated financial losses of building functions due to a building system failure divided by the frequency (MTBF) of the failure.

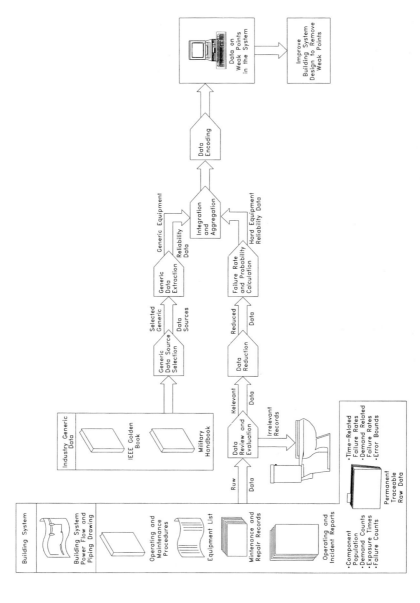

Figure 17.1 The RAM (reliability, availability, and maintainability) database development process for industrial and commercial facilities.

Availability: A ratio that describes the percentage of time a system component or system can perform its required function.

Component: A piece of electrical or mechanical equipment, an interconnecting pipe/valve or circuit, a section of building system, or a group of items that is viewed as an entity for the purpose of reliability evaluation.

Failure: The termination of the ability of an component to perform a required function.

Failure Rate: The mean number of failures of a component per unit exposure time.

Forced Downtime: The average time per year a system or equipment is unavailable in between failures and expressed in hours per year.

Lambda (λ): The inverse of the mean exposure time between consecutive failures. λ is typically expressed in failures per year.

MTBF: The mean exposure time between consecutive failures of a component or system. The mean time between failures is usually expressed in years per failure. For some applications, measurement of mean time between repairs (MTBR) rather than mean time between failures may provide more statistically correct information.

MTTR: The mean time to repair a failed component. For a system, it is the total amount of time it is unavailable in between failures and is expressed in hours in both cases.

Point: Any place or location within the building critical system. The name or designation for a point is always the same as the name of the zone in which the point is located.

RAM Table: A lookup historical and field data table that displays the MTBF and MTTR for system components.

Reliability: An indication of the ability of a component or system to perform its intended function during a specified time.

Restore Time: The time to restore is the sum of the mean time to repair (MTTR) for the failure plus the computed time to restream or restart the connected load.

Building System: A group of components connected or associated in a fixed configuration to perform a specified function. Function could be distributing power or air conditioning, circulating water, controlling system interoperation, etc.

Zone: A zone is defined as a segment of a power (or a segment of water or steam distribution system) in which a fault (or leak) at any location within the segment or zone would have the common impact of causing the first upstream protective device to isolate the system (or in case of water/steam create mess and empty the system).

17.3 Building System Reliability Analysis Procedure

Failure rates for series and parallel systems combine in a manner very similar to that of impedances. The only problem with this approach is that while an array of electrical/mechanical components may be physically connected in parallel, its reliability equivalent model may have the components connected in series. For example, if there is a district chilled water line running along several buildings with tap offs into each building, the flow diagram would show all the building tap offs in parallel to the central chiller plant. Yet when one building system experiences a major leak, all of the other buildings also lose flow until the problem can be isolated and repaired. Thus, when representing the reliability of all the tap offs, they end up being in series. To overcome this misrepresentation the distribution system should be divided into zones.

The concept of developing reliability zones is quite similar to the technique used in protective relaying for describing zones of protection. One of the primary purposes of protective devices from the standpoint of a system is to improve reliability by isolating faulty components while minimizing the impact to the remainder of the system. When configuring a system, a zone can be configured either in series with a single source point immediately above it, or a zone can be configured in parallel with the two sources immediately above it.

The analysis should determine the failure rate and repair rate of all components within a zone without taking into consideration that the zone is connected to any other portion of system. The analysis should then determine the failure rate and repair rate for all the points in the system by taking into consideration the configuration of all the zones. Create a matrix called the zone table in which the configuration of the zones can be entered. Create another matrix table, called the unit impact table, which allows the entry of various interconnected units or loads and the impact that each point has on the unit or load. It should also include data entry to reflect the cost impact for not having the unit available for 24 h. The last data entry field should be the time it will take to get the unit or load reenergized from a cold start.

The analysis should determine if the unit has been down long enough to become "cold." For most critical application units, the time required to restream the unit following a failure is relatively short if the problem can be corrected quickly. However, if the repair persists for an extended period of time, the unit will get "cold" and restreaming of the unit may take much longer, The system analysis should have a result section labeled the "Critical Building System Component Summary." Here the total quantity of each system component is displayed along with the expected failure rate for that population of components. The summary should include the total number of components within the critical system analyzed and the failure statistics for the entire population of components.

The following is a brief description of the computations normally used when combining zones in series and parallel.

17.3.1   Component Analysis

A. Single Component Analysis

Reliability and availability are necessary to describe the characteristics of the single component shown (see Figure 17.2a).

$$d_1 = \text{MTBF (in hours)}$$

$$r_1 = \text{MTTR (in hours)}.$$

Reliability of components is frequently given as failures per year of operating time. Using these numbers, the MTBF can be calculated using

$$d_1 = \text{MTBF (in hours)}$$

$$d_1 = 8760 \text{ h/(failures/year)}.$$

The failure rate (λ) is given by

$$\lambda = 1/\text{MTBF}$$

and the component reliability (R_1) for one year is given by

$$R_1 = e^{-(\lambda)(8760)}.$$

For a single component, the availability (A) is given as the total operating time over the total time, or

$$A = (\text{MTBF})/(\text{MTBF} + \text{MTTR}).$$

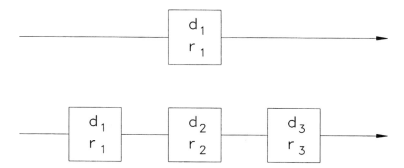

Figure 17.2(a)   System with components in series.

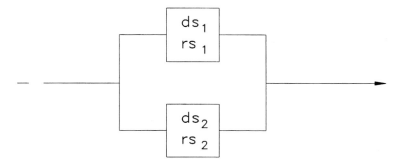

Figure 17.2(b)   Redundant system with components in parallel.

B. Systems with Components in Series

For the system shown with three different components in series (see Figure 17.2b)

$$d = \text{MTBF (in hours)}$$

$$r = \text{MTTR (in hours).}$$

The characteristics of each individual component can be calculated using

$$\lambda_i = 1/d_i$$

$$A_i = d_i/(d_i + r_i)$$

$$R_i = e^{-\lambda_i t}$$

$$R_i = e^{-\lambda_i (8760)} \text{ for one year.}$$

Using these, the combined failure rate (failures per year) becomes

$$\lambda_s = \lambda_1 + \lambda_2 + \lambda_3$$

or

$$\lambda = R_1 * R_2 * R_3.$$

Reliability of the system for one year is

$$R_s = e^{-\lambda_s (8760)}.$$

System availability is

$$A_s = A_1 * A_2 * A_3.$$

Probability of failure during one year is

$$P_s = (1 - R_s) * 100.$$

MTTR in hours

$$\text{MTTR} = \text{MTBF}[(1/A_s) - 1].$$

where $\text{MTBF} = 1/\lambda_s$
  Forced unavailable time (FUT) is

$$\text{FUT} = (1 - A_s).$$

C. Systems with Redundant Components

1. *Parallel (redundant) systems:* Reliability can be dramatically increased by installing a parallel (redundant) system. The simplest of these is a system that operates satisfactorily if either one of two parallel components functions (see Figure 17.2c).

The reliability for such a system for one year can be calculated using

$$R_s = [1 - (1 - R_{s1})(1 - R_{s2})].$$

The combined failure rate is

$$\lambda_s = \ln(1/R_s).$$

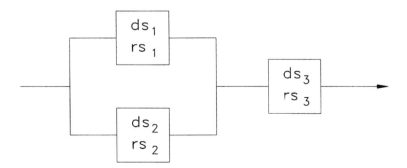

Figure 17.2   (c) System with components in series with redundant system.

The system availability is

$$A_s = [1 - (1 - A_{s1})(1 - A_{s2})]$$

and the probability of failure during one year is

$$P_s = (1 - R_s) * 100.$$

2. *Parallel (redundant) systems with maintenance provisions:* If the components can be repaired, the reliability of the systems described above also becomes a function of the time required to repair the system.

Using constant failure rate for two identical units, the steady-state availability is

$$A = \mu/\lambda + \mu$$

where the repair rate (μ) is [(MTTR) $-$ 1] and the MTBF is

$$MTBF = \mu/(2\lambda^2).$$

For two different components in parallel, the MTBF is

$$MTBF = (\lambda_a + \mu_b)(\lambda_b + \mu_a) + \lambda_a(\lambda_a + \mu_b)$$
$$+ (\lambda_b + \mu_a)/\lambda_a\lambda_b(\lambda_a + \lambda_b + \mu_a + \mu_b)$$

and the steady-state availability for *n* blocks is

$$A = 1 - \prod_{i=1}^{n}(\lambda_i/\lambda_i + \mu_i)$$

or for two parallel components

$$A = 1 - [(\lambda_1/\lambda_1 + \mu_1))(\lambda_2/(\lambda_2 + \mu_2))].$$

D. Common Cause

Redundancy calculations frequently lead to reliability numbers that are outside the realm of reason, for example, one failure in a thousand years. In reality, even redundancy of components still leaves a chance that the parallel system will fail from a common mode. Examples of this include common electrical connections, common control wiring, or the chilled water piping. One suggested method to account for common failure modes is to use an additional component in series with the redundant system as shown in Figure 17.2c. A reliability of between 0.01 and 0.1 times the reliability of the components in the individual redundant components seems to provide results that are comparable with actual experiences.

17.4 Risk Analysis

Risk analysis is very important in considering what are called the failure modes. Determine first what is a failure and second, the consequence of the failure. Most of the time it is more important to know all the failure possibilities than to have an in-depth knowledge of some and miss others. This first step is to find out what could happen, not how often or how severe.

The use of an event tree is most helpful. Particular care must be given to find "common mode" causes. Common mode failures are those single events that can cause several units to fall. A main power switch is a typical common mode failure because it may disconnect several units where an individual circuit fuse will cut out only an individual unit.

Risk is defined differently by an engineer than a by layman. Table 17.1 indicates salient differences between the two.

The primary goal of any building system reliability analysis is to reduce the probability of accidents and the attending human, economic, and environmental losses. The human losses include death, injury, and sickness or disability and the economic losses include production, revenue, or service shutdowns; loss of capital equipment; legal costs; and regulatory agency fines. Environmental losses are air and water pollution and other environmental degradations such as odor, vibration, and noise.

Failures occur when an initiating event is followed by system failures. The three types of basic failure causes are:

1. Humans: operator error, design error, and maintenance error
2. Hardware failure: power source component failure, leakage of toxic refrigerants from a valve, loss of motor lubrication, incorrect sensor measurement, controller communication failure; etc.
3. Environmental disaster: earthquake, storm, flood, tornado, lightning; etc.

Disastrous failures are frequently caused by a combination of failure events, that is, a hardware failure plus human error and/or environmental faults. Typical policies to minimize these accidents include:

1. Equipment redundancies
2. Energy transportation media and source backups
3. Inspection and maintenance
4. Safety systems such as sprinklers, fire walls, and relief valves
5. Fail-safe and fail-soft design

Table 17.1 Risks and Difference in Technical Treatment and Layman Treatment

Approach	Technical Treatment	Layman Treatment
1. Risk acceptance	Emphasizes sound decision making, because of finite resources for risk reduction and impractability of achieving zero risk; checks the tendency to ignore nondollar costs in such tradeoffs and provides net benefit to system; neglects indirect and certain long-term benefits without explicit decision criteria and structured analyses.	Greater tendency to judge risk in absolute terms; ignores risks of no-action alternatives to rejected technology; gives greater weight to nondollar costs; emphasizes personal rather than system benefits; includes both qualitative and quantitative benefits but tends to neglect indirect and long-term benefits; tends to distort equity considerations in favor of personal interest.
2. Risk-assessment methods	Quantitative, experimental, computational • Risk = consequence × probability • Fault trees/event trees • Statistical calculation • Engineering test equipment and simulators	Qualitative, intuitive, impressionistic • Incomplete rationale • Emotional input to value judgments • Personal experience/memory • Media accounts • Political exchange
3. Basis of information	Established institutions and through qualification of experts	Nonestablishment sources; limited ability to judge qualifications or nonpractical engineering knowledge
4. Risk-attribute evaluation	Provides objective, conservative assessment; considers broad range of high and low estimates and gives equal weight to diverse views over treatment of incommensurables and discount rate.	Tends to exaggerate or ignore risk; gives greater weight to catastrophic and immediate issue except for overall exposure to failure; gives greater weight to dreaded risk.

17.5 Failures and Relationships

A system consists of hardware, materials, and plant personnel, is surrounded by its physical and social environment, and suffers from aging (wearout or random failure). Accidents are caused by one or a set of physical components generating failure events. Each physical component in an integrated system is related to the other components in a specific manner. The system environment, in principle, includes the entire world outside the plant, so an appropriate boundary for the environment is necessary to prevent the initiating-event and event-tree analyses from diverging.

A primary probability risk analysis objective is to identify the causal relationships between human, hardware, and environmental events that result in failures or accidents, and to find ways of ameliorating their impact by building system redesign and upgrades. The causal relationships can be developed by event and fault trees, which are analyzed both qualitatively and quantitatively. After the combination of the basic failure events that lead to losses are identified the building system can be improved and losses reduced.

17.6 Fault Trees

The fault tree is structured such that an undesired system failure event appears at the top and is linked to more basic failure events by event statements and logic gates. The fault tree analysis is restricted only to the identification of the system and component causes that lead to one particular top event. System specification requires a careful delineation of component initial conditions. All components that have more than one operating state generate initial conditions. For example, if the initial quantity of fluid in a tank is unspecified, the event "tank is full" is one initial condition, while "tank is empty" is another.

The fault-tree structure provides following:

1. Reveals system weakest links.
2. Points out the segments of the building system important to the system failure of interest.
3. Provides a graphic aid.
4. Provides options for qualitative and quantitative system reliability analysis.
5. Allows concentration on one particular system failure at a time.
6. Provides an insight into system behavior.

Fault-Tree Construction: In large fault trees, mistakes are difficult to find, and the logic is difficult to follow or obscured. For large systems it is advised to divide the system into logical boundaries. The construction of fault trees is perhaps as much of an art as a science. Fault-tree structures are not unique; no two analysts construct identical fault trees (although the trees should be equivalent in the sense that they yield the same cut set or combination of causes).

To find and visualize causal relations by fault trees, gate symbols and event symbols are used as main building blocks. Gate symbols include AND, OR, inhibit, priority AND, exclusive OR, and voting. Event symbols are rectangle, circle, diamond, house, and triangle. The symbols for the gates are listed in Figure 17.3. A gate may have one or more input events but only one output event.

Gate Symbols: Connect events according to their causal relationships. The *AND gate* output event occurs if all input events occur simultaneously, and the *OR gate* output event happens if any one of the input events occurs. Examples of

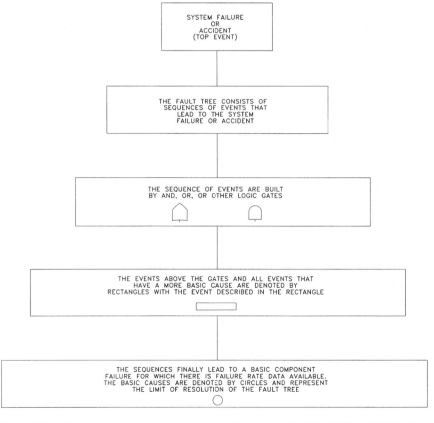

Figure 17.3   Fundamental fault-tree structure. (Copyright © 1996. IEEE. All Rights Reserved)

OR and AND gates are shown in Figure 17.4. For example consider a fuel pumping system shown in Figure 17.5, the system event "pressure tank overflows" happens when two events, "pump motor runs too long" and "overflow protection device does not operate," occur simultaneously. The latter event happens when either one of the two events, "high level cutoff protective device does not exist" or "device is not suitable," occurs. Showing these events as rectangles implies they are system states. If the event "overflow device malfunction," for example, were a basic cause it would be circled and become a basic hardware failure event. The causal relationship expressed by an AND gate or OR gate is deterministic because the occurrence of the output event is controlled by the input events.

There are causal relationships that are not deterministic. Consider the two events "overflow occurs" and "protective device fail to operate." The causal relationship here is probabilistic, not deterministic, because a protective device misoperation does not always result in overflow. The *hexagonal inhibit gate* is used to represent a probabilistic causal relation. The event at the bottom of the inhibit gate is an input event, whereas the event to the side of the gate is a conditional event. The output event occurs if both the input event and the conditional event occur. In other words, the input event causes the output event with the (usually constant, time-independent) probability of occurrence of the conditional event. In contrast to the probability of equipment failure, which is usually time dependent, the inhibit gate frequently appears when an event occurs with a probability according to a demand.

Another gate logically equivalent to an AND gate is the *priority AND gate.* It causes the output if input events occur in a specific order. The occurrence of the input events in a different order does not cause the output event. Consider, for example, a system that has a principal power supply and a standby power supply. The standby power supply is switched into operation by an automatic switch when the principal power supply fails. Power is unavailable in the system if the switch controller fails first and then the principal unit fails. The failure of the switch controller after the failure of the principal unit does not yield a loss of power because the standby unit has been switched correctly.

Similarly *Exclusive OR gates* describe a situation where the output event occurs if either one, not both, of the two input events occurs. Consider a system powered by two generators. A partial loss of power can be represented by the exclusive OR gate. In the case of multiple generators a *Voting gate* based on the occurrence of *m*-out-of-*n* input events represents a partial loss. This situation can be expressed by the m-out-of-n gate.

Event Symbols: Describe occurrence (see Figure 17.6). In the fault tree schematics, a *rectangle* usually denotes an undesirable system event, resulting from basic failures through logic gates. The *circle* designates a basic component failure that represents the lowest level of a fault tree. Circles usually represent events (components) for which failure rate (likelihood of occurrence) data are available. "Pump fails to start," "pump fails to run," or "pump is out for mainte-

	GATE SYMBOL	GATE NAME	CAUSAL RELATION
1		AND GATE	OUTPUT EVENT OCCURS IF ALL INPUT EVENTS OCCUR SIMULTANEOUSLY
2	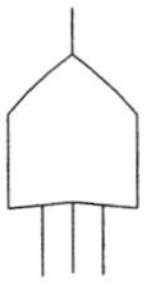	OR GATE	OUTPUT EVENT OCCURS IF ANY ONE OF THE INPUT EVENTS OCCURS
3	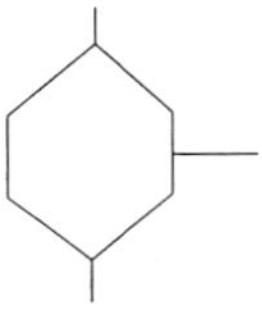	INHIBIT GATE	INPUT PRODUCES OUTPUT WHEN CONDITIONAL EVENT OCCURS
4	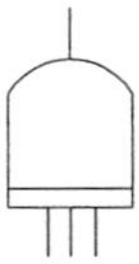	PRIORITY AND GATE	OUTPUT EVENT OCCURS IF ALL INPUT EVENTS OCCUR IN THE ORDER FROM LEFT TO RIGHT.
5		EXCLUSIVE OR GATE	OUTPUT EVENT OCCURS IF ONE, BUT NOT BOTH, OF THE INPUT EVENTS OCCUR
6	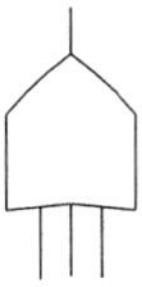	m-OUT-OF-n GATE (VOTING OR SAMPLE GATE)	OUTPUT EVENT OCCURS IF m-OUT-OF-n INPUT EVENTS OCCUR.

Figure 17.4 Gate symbols. (Copyright © 1996. IEEE. All Rights Reserved)

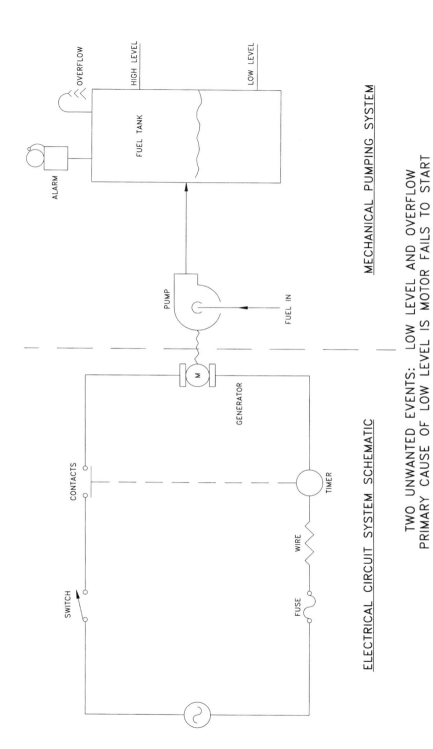

ELECTRICAL CIRCUIT SYSTEM SCHEMATIC

MECHANICAL PUMPING SYSTEM

TWO UNWANTED EVENTS: LOW LEVEL AND OVERFLOW
PRIMARY CAUSE OF LOW LEVEL IS MOTOR FAILS TO START

Figure 17.5 Fuel pumping system: electrical/mechanical schematic diagram.

	EVENT SYMBOL	MEANING OF SYMBOL
1	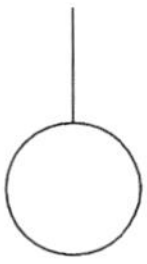 CIRCLE	BASIC COMPONENT FAILURE EVENT WITH SUFFICIENT DATA
2	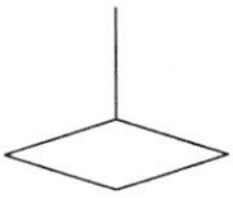 DIAMOND	UNDEVELOPED EVENT
3	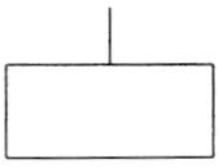 RECTANGLE	STATE OF SYSTEM OR COMPONENT EVENT
4	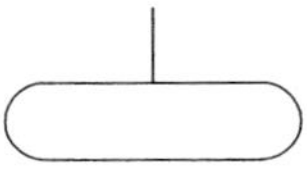 OVAL	CONDITIONAL EVENT WITH INHIBIT GATE
5	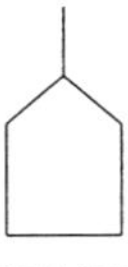 HOUSE	HOUSE EVENT. EITHER OCCURRING OR NOT OCCURRING
6	 TRIANGLES	TRANSFER SYMBOL

Figure 17.6   Event symbols. (Copyright © 1996. IEEE. All Rights Reserved)

nance" are examples of basic component failures represented by a circle. *Diamonds* are used to symbolize undeveloped events, in the sense that a detailed analysis into the basic failures is not carried out because of lack of information, money, or time. "Failure due to sabotage" is an example of an undeveloped event. Such events are removed frequently prior to a quantitative analysis. They are included initially because a fault tree is a communication tool and their presence serves as a reminder of the depth and bounds of the analysis. Most secondary failures are diamond events.

To examine various special fault-tree cases by forcing some events to occur and other events not to occur a *house* symbol is used. When we turn on the house event, the fault tree presumes the occurrence of the event and vice verse when we turn it off. *Triangles* are used to cross-reference two identical parts of the causal relationships. The two triangles have the same identification number. The transfer-out triangle has a line to its side from a gate, whereas the transfer-in triangle has a line from its apex to another gate.

See Figure 17.7 to construct a fault tree for the integrated system shown in Figure 17.5. This tree construction uses component failure characteristics shown in Figure 17.8.

17.7 Intelligent Building Automation Reliability

The term fault-tolerant intelligence computing can be defined as "the ability to execute specified algorithms correctly regardless of hardware errors and program errors." The understanding of fault tolerance can be helped first by understanding building intelligent system reliability, data protection, and maintainability needs. Faults in intelligence systems can be defined as "the deviation of one or more logic variables in the computer hardware from their design-specified values." A logic value for a digital computer is either a zero or a one. A fault is the appearance of an incorrect value such as a logic gate "stuck on zero" or "stuck on one." The fault causes an "error" if it in turn produces an incorrect operation of correctly functioning logic elements. Therefore, the term "fault" is restricted only to the actual hardware that fails.

Faults can be classified in several ways. Their most important characteristic is a function of their duration. They can be either permanent (solid) or transient (intermittent). Permanent faults are caused by solid failures of components. They are easier to diagnose, but usually require the use of more drastic correction techniques than do transient faults. Transient faults cause 80% to 90% of faults in most systems. Transient faults or intermittents can be defined as random failures that prevent the proper operation of a unit for only a short period of time, not long enough to be tested and to become permanent. Permanent fault-tolerant techniques must then be used for system recovery.

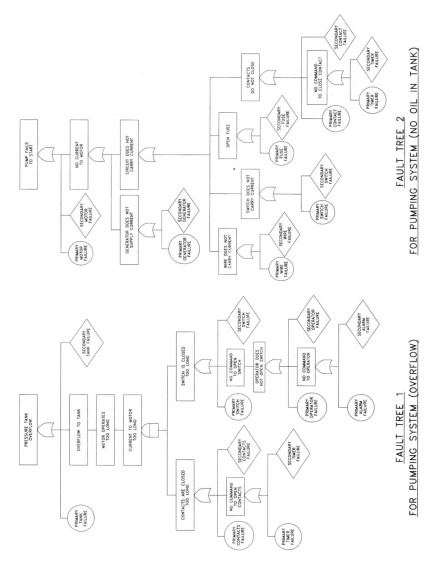

Figure 17.7 Fuel pumping system—Fault trees.

270

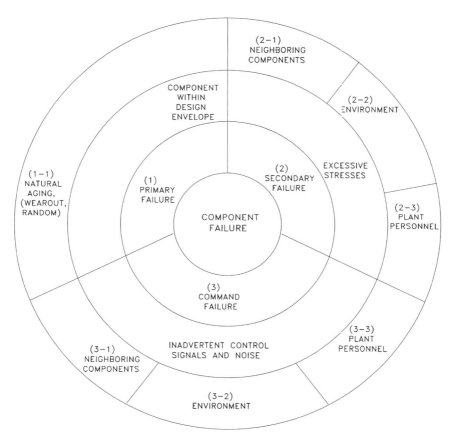

Figure 17.8 Component failure characteristics. (Copyright © 1996. IEEE. All Rights Reserved)

17.7.1 Computing Reliability

The goal of systems reliability or of fault-tolerant computing is either to prevent or allow recovery from faults and continue correct system operation. This also includes immunity to software faults induced into the system. To achieve high reliability, it is essential that component reliability be as high as possible. As the complexity of computer systems increases, almost any level of guaranteed reliability of individual elements becomes insufficient to provide a satisfactory probability of successful task completion. Therefore, successful fault-tolerant computers must use a judicious selection of protective redundancy to help meet the reliability requirements. The three redundancy techniques are:

1. Hardware redundancy
2. Software redundancy
3. Time redundancy.

These three techniques cover all methods of fault tolerance. Hardware redundancy can be defined as any circuitry in the system not necessary for normal computer operation should no faults occur. Software redundancy, similarly, is additional program instructions present solely to handle faults. Any retrial of instructions is also known as time redundancy.

17.7.2 Hardware Redundancy

This is the set of all of those hardware components that need to be introduced into the system to provide fault tolerance with respect to operational faults. These components would be superfluous should no faults occur, and their removal would not diminish the computing power of the system in the absence of faults.

In achieving hardware fault tolerance, it is clear that one should use the most reliable components available. However, increasing component reliability has only a small impact on increasing system reliability. Therefore, it is more important to be able to recover from failures than to prevent them. Redundant techniques allow recovery, and thus are very important in achieving fault-tolerant systems. The techniques used in achieving hardware redundancy can be divided into two categories: static or masking redundancy and dynamic redundancy.

Static techniques are equally effective in handling both transient and permanent failures. Many different techniques of static redundancy can be applied. The simplest or lowest level of complexity is by a massive replication of the individual components of the system. For example, four diodes connected as two parallel pairs that are themselves connected in series will not fail if any one diode fails "open" or "short." Logical gates in similar arrangements can also guard against single faults and even some multiple faults for largely replicated systems.

More sophisticated systems use replication at higher levels of complexity to mask failures. Masking is virtually instantaneous and automatic. It can be defined as any computer error correction method that is transparent to the user and often to the software. Redundant components serve to mask the effect of hardware failures of other components. Instead of using a mere massive replication of components configured in fault-tolerant arrangements, identical nonredundant computer sections or modules can be replicated and their outputs voted upon. Examples are triple modular redundancy (TMR) and more massive, N-modular redundancy (NMR) where N can stand for any odd number of modules.

In addition to component replication, coding can be used to mask faults as well as to detect them. With the use of some codes, data that have been garbled, that is, bits changed owing to hardware errors, can sometimes be recovered instantaneously with the use of redundant hardware. Dynamic recovery methods are, however, better able to handle many of these faults.

Higher levels of fault tolerance can be achieved more easily through dynamic redundancy, implemented through the dual actions of fault detection and recovery. This often requires software help in conjunction with hardware redundancy. Many dynamic methods are extensions of static techniques.

More effective use of massive redundancy in components often results from dynamic control. Redundant modules or spares can have a better fault tolerance when left unpowered until needed, because they will not degrade while awaiting use. This technique—stand-by redundancy—can use dynamic voting techniques to achieve a high degree of fault tolerance. The union of dynamic and stand-by redundancy is referred to as hybrid redundancy. Hybrid redundancy requires additional hardware to detect and switch out faulty modules and switch in good spares.

Error detecting and error correcting codes can be used dynamically to achieve fault tolerance in a computing system. Coding refers to the addition and rearrangement of the bits in a binary word that contains information. The strategy of coding is to add a minimum number of check bits (the additional bits) to the message in such a way that a known degree of error detection or correction is achieved. Error detection and correction is accomplished by comparing the new word, which should be unchanged after transmission, storage, or processing, with a set of allowable configurations of bits. Discrepancies thus discovered signal the existence of a fault that then can be corrected if enough of the original information remains intact—that is, the original binary word can be reconstructed provided a set number of bits in the coded word have not changed. Encoding and decoding words with the use of redundant hardware can be very effective in detecting errors. Often, through hardware or software algorithms, incorrect data also can be reconstructed. Otherwise, the detected errors can be handled by module replacement and software recovery actions. The actions taken are dependent on the extent of the fault and of the recovery mechanisms available to the computing system.

17.7.3 Software Redundancy

Redundant software plays a major role in most fault-tolerant computers. Even computers that recover from failures mainly by means of hardware use software to control their recovery and decision making processes. The level of software used depends upon recovery system design, and recovery design is dependent upon the expected type of error or malfunction. Different schemes have been found to be appropriate for the handling of different errors. Some types of error correction are most efficiently accomplished with hardware. Others need only software, but most use a mixture of the two.

For a functional system, that is, one without hardware design faults, errors can be classified into two varieties:

1. Software design errors
2. Hardware malfunctions.

The first category can be corrected mainly by means of software. It is extremely difficult for hardware to be designed to correct for programmer's errors. Software methods, though, are often used to correct hardware faults, especially transient ones. There are several software errors that computers may be designed to detect, including the use of illegal instructions (i.e., instructions that do not exist), the use of privileged instructions without authorization, and address violations (i.e., reading or writing into locations that are beyond those of usable memory). These limits can often be set physically on the hardware. Computers capable of error detection cause interrupts, which route the program to specific locations in memory. The programmer, knowing these locations, can add his own code to branch to his specific subroutines which can handle each error specifically as he sees fit.

There are several methods of software recovery from software errors. As mentioned before, parallel programming where alternative methods are used to determine a correct solution can be used when an incorrect solution can be determined. Some less sophisticated systems print out diagnostics so that the user can correct the program off-machine. This should only be a last resort for a fault-tolerant machine. Nevertheless, a computer should always keep a log of all errors incurred, memory size permitting.

Preventive measures used with software methods refer mainly to the use of redundant storage. Hardware failures often result in a garbling or a loss of data or instructions read from memory. If hardware techniques such as coding cannot recover the correct bit pattern, those words will become permanently lost. Therefore, it is important to at least duplicate all necessary program and data storage so that it can be retrieved in case one copy is destroyed. In addition, special measures should be taken so that critical programs such as error recovery programs are placed in nonvolatile storage, that is, read only memory. Critical data also should be placed in nondestructive readout memories, such as plated-wire memories.

The second task of the software in fault tolerance is to be able to detect and diagnose errors. Software error detection techniques often can be used to detect hardware faults that are transient. This is important since "a relatively large number of malfunctions are intermittent in nature rather than solid failures." Time-redundant processes, that is, repeated trials, shall be used for their recovery.

Software error detection techniques do not locate the sources of the errors. Therefore, diagnostic test programs are often used to locate the module or modules responsible. These programs usually test the extent of the faults at the time

of failure or do periodic tests to determine malfunctions before they manifest themselves as errors during program execution. Almost every computer system uses some form of diagnostic routines to locate faults. In a fault-tolerant system, the system itself initiates these tests and interprets their results, rather than requiring the outside insertion of test programs by operators.

18

Air, Water, and Power Quality and Building Systems

Consider a traditional building. Yet in it: You provide security locks to the building entrance as filters to pass certain humans and not others. You provide water filters to avoid microorganisms and bacteria in drinking water. You provide electrical filters as barriers to unwanted frequency and power distortion. You provide air filters, thus supplying air is dirt filtered, heated, or cooled. You provide vapor filters against leakage moisture transmission. All these basic filters basically provide clean air, power, and water to building habitats. In a high tech building this requirement is expanded to microlevels and expectation of building systems are at that level.

18.1 Indoor Air Quality (IAQ) and Modern Buildings

Airborne materials include gases and particulates that may be generated by occupants and their activities in a space, from outgassing and/or shedding of building materials and systems, originate in outside air, and/or created from building operating and maintenance programs and procedures. Refer back to Figure 14.6 showing sources of air impurities and configuration of building system to clean impure air similar to that of building habitats. Airborne particles include bioaerosols, asbestos, man-made mineral fibers, and silica, and gaseous contaminants include radon and soil gases as well as volatile organic compounds.

18.1.1 Contaminants and Their Sources

Bioaerosols: These are airborne microbiological particulate matter derived from viruses, bacteria, fungi, protozoa, mites, pollen, and their cellular or cell mass components. Bioaerosols are everywhere in the indoor and outdoor environments.

Some common bioaerosoles sources are saprophytic bacteria and fungi in the soil and in the atmosphere, for example, *Cladosporium,* a fungus commonly found on dead vegetation and almost always found in outdoor air. *Cladosporium* spores are also found in indoor air, depending on the amount of outdoor air that infiltrates into interior spaces or is brought into the HVAC system. Bacteria that are saprophytic on human skin (*Staphylococcus*) and viruses (e.g., Influenza A) that are parasites in the human respiratory tract are shed from humans and are thus commonly present in indoor environments.

Although microorganisms are normally present in indoor environments, the presence of abundant moisture and nutrients in interior niches amplifies the growth of some microbial agents to the extent that the interior environment is microbiologically rich. Thus, certain types of humidifiers, water spray systems, and wet porous surfaces can be reservoirs and sites for growth of fungi, bacteria, protozoa, or even parasites. Turbulence associated with the start-up of air-handling unit plenums may also elevate concentrations of bacteria and fungi in occupied spaces.

The presence of bioaerosoles or microorganisms in indoor environments may cause infective and/or allergic building-related illnesses. Some microorganisms under certain conditions may produce bad odors and volatile chemicals that are irritative, thus contributing to the development of what is called the *sick building syndrome.* Outbreaks of infectious illness in the indoor air may be caused by other types of microorganisms, such as viruses from ill persons.

Asbestos: The term asbestos refers to a group of naturally occurring silicates that occur in fiber bundles with unusual tensile strength and fire resistance. In older buildings they are found in insulation products. This material is banned in building materials. In old buildings, exposure can occur owing to natural degradation, renovation, and removal.

A variety of diseases have been associated with asbestos exposure, such as asbestosis (a chronic disease of the lung tissue leading to respiratory failure because of inadequate ability to transport oxygen across the lung tissue) and cancers of the lungs, larynx, and gastrointestinal tract (esophagus, stomach, colon, and rectum).

Asbestos use in the United States is strictly regulated through federal agencies. Occupational Safety and Health Administration (OSHA) has jurisdiction in the workplace, with an overall permissible exposure limit set the same for both general industry and construction. The Environmental Protection Agency (EPA) regulates asbestos use and exposure in a variety of other settings, including schools and residential environments. Strict guidelines are provided, including a document on safe handling and abatement/remediation.

Man-Made Mineral Fibers (MMMFs): Also known as man-made vitreous fibers (MMVF), these are synthetic, amorphous, noncrystalline vitreous structures. They are used in buildings for insulation materials. Since they are amorphous, they do not split lengthwise, as do crystalline fibers such as asbestos.

They break across their diameter, leading to progressively shorter pieces. Under the influence of high temperatures, refractory fibers may change their chemical/physical characteristics to crystalline silica.

Possible health effects from man-made mineral fibers are still very controversial. Broad categories of possible health effect are: *Dermatitis*—itching and erythema (reddening of skin) occur after dermal exposure. *Microbial contamination*—exposure to specific microbial agents may sensitize some individuals. *Lung disease*—several surveys have examined the relationship between MMMF exposure and interstitial lung disease, the kind of disease associated with asbestos exposure. At present, there is no evidence that MMMF causes interstitial lung disease. *Cancer and mesothelioma*—man-made mineral fibers have been suspected of leading to elevated cancer rates. Still, at present there is no clear evidence that any forms of MMMF are carcinogenic in humans. However, the International Agency for Research on Cancer and the EPA have classified mineral wool and rock wool as possible human carcinogens.

OSHA regulates man-made mineral fibers as nuisance dust, at 15 mg/m^3 for the construction industry. National Institute for Occupational Safety and Health (NIOSH) proposes a recommended exposure level of 3 f/ml (fibers/ml) for respirable fibers and a level of 5 mg/m^3 for total fibrous glass. The American Conference of Governmental Industrial Hygienists (ACGIH) recommends 10 mg/m^3. Sweden has set a level of 1 f/mL for respirable fibers, and Denmark has set a level of 2 f/mL.

Silica: Silica consists of silicon dioxide, which is abundant in the earth's crust and is used in the manufacture of glass, refractories, abrasives, buffing and scouring compounds, and lubricants. Its biological effects depend on the grade of the silica. Silica is associated with three forms of silicosis—acute, accelerated, and chronic (nodular). They are characterized by an immunologic action of silica in the lungs, leading to progressive tissue destruction and immunological abnormalities. Standards have been established separately by OSHA and Mine Safety and Health Administration (MSHA). The environmental criterion is five (5) fibers per milliliter. Sampling may occur on particulate filters. Quantification is through low-temperature ashing or X-ray diffraction spectrometry.

Radon and Other Soil Gases: Radon is a naturally occurring, chemically inert, odorless, tasteless radioactive gas. It is produced from the radioactive decay of radium, which is formed through several intermediate steps from the decay of uranium. Additional, but secondary, sources of indoor radon include groundwater and radium-containing building materials.

Radon gas enters buildings through cracks or openings such as sewer pipe and sump pump openings, cracks in concrete, and wall–floor joints. The amount of radon entering and subsequent indoor concentration distribution depends on several factors, including the concentration of radium in the surrounding soil or rock, the soil porosity and permeability, and the air pressure differential between the building and the soil or between various indoor spaces which may result from the stack effect, operation of exhaust fans, or operation (or lack of) HVAC equipment. While several sources of radon may contribute to average radon levels in buildings, pressure-driven flow of soil gas constitutes the principal source.

The health effects associated with radon are mainly due to exposures to radon decay products, and the amount of risk is assumed to be directly related to the total exposure. Indoor concentrations of radon can vary hourly, daily, and seasonally, in some cases by as much as a factor of 10 to 20 on a daily basis. For example, measurements made during a mild spring may underestimate the annual average level because of ventilation from open windows or the operation of HVAC equipment (e.g., economizer operation, which increases outdoor air ventilation rates). Similarly, indoor levels during a cold winter may be higher than average because the building is sealed and the outdoor air ventilation rates are minimized. Thus, longer-term measurements (6 months to 1 year) made during normal use generally provide more reliable estimates of the average indoor concentration.

Exposure to indoor radon may be reduced by preventing radon entry, or removing or diluting radon or radon decay products after entry. To prevent entry, sealing various parts of the building substructure has been attempted, but it is often not completely effective by itself, since some openings in the building shell may not be accessible or new openings can develop with time. Because furnaces or HVAC systems can contribute to the depressurization of buildings, the location and tightness of supply and return ducts are important, especially in new construction.

Any volatile organic compounds present in soil gases in the vicinity of buildings may enter those buildings under certain, as yet undefined, soil and building conditions. Therefore, their presence, or possible presence, should be considered in the design, construction, and operation of buildings—particularly in the vicinity of inactive landfills or hazardous waste sites. Buildings in these areas should be made to minimize substructure leakage, to prevent substructures from being depressurized by the operation of HVAC systems, and to seal duct systems in crawl spaces or beneath slab foundations against the possible intrusion of gases. Techniques installed to prevent entry of radon gas will also prevent entry of other soil gases.

Volatile Organic Compounds (VOCs): These are air pollutants found in all nonindustrial environments. After ventilation, VOCs are probably the first concern when diagnosing an IAQ problem. VOCs are organic compounds with vapor pressures greater than about 10^{-3} to 10^{-4} mm Hg (torr). While building and furniture materials are known to emit VOCs, ventilation may also transport outdoor pollutants to the indoor environment; so the ventilation system itself may be a source of VOCs. Refer back to table 4.1 for indoor air quality safe limits of some gases and chemicals naturally present in ambient air.

Adverse health effects potentially caused by VOCs in buildings indoor environments fall into three categories: (1) irritant effects, including perception of unpleasant odors and mucous membrane irritation; (2) systemic effects, such as fatigue and difficulty concentrating; and (3) toxic chronic effects, such as carcinogenicity. Approaches to reducing indoor exposures to VOCs include the following:

- Use low-emitting products indoors.
- Increase general ventilation, although this may not be energy-efficient and may not be effective for some sources, such as building materials, where increasing the ventilation six times decreases the VOC only by 50%.
- Install local ventilation, that is, local exhaust ventilation, near photocopiers, printers, and other point sources.

Odors: Good or bad odors are caused by molecules released from the surface of various substances. A fresh fish will have hardly any smell at all, but bacteria on a decomposing fish will release strong odor molecules. The olfactory receptors in the nose will immediately detect even a few of these molecules. Odors can be eliminated physically, chemically, or biologically. Removing odors physically can be accomplished with activated carbon filters, as in the air handling units. A tiny crystal of activated carbon contains countless holes, or pores, that give the crystal a large surface area relative to its volume. Odor molecules floating on air currents collide with the carbon particles and become trapped in these pores. Other agents that work in a similar way include silica gel and activated bauxite.

Odors can be chemically removed by using acids to neutralize alkaline odors and alkalies to neutralize acid odors. But only a limited number of substances can be treated chemically.

For biological odor removal, microorganisms are used to break down odor molecules, rendering them odorless. Biological agents are limited, however, by conditions such as temperature. Because of such limitations, activated carbon is the most effective means of odor removal.

Pollens: These are allergens for a substantial fraction of the population. It may not be such a good idea for agriculture and horticulture to eliminate them, but removal in indoor spaces in buildings is desired. *Ventilation* involves the provision of "fresh air" to interiors to replenish the oxygen used by people and to help carry away their by-products of carbon dioxide and bodily odors. Ventilation is desirable all year round; local and national codes of advanced countries recommend minimum circulation rates of fresh air required based on building occupancy and functions.

Air Handling Systems: These are not only potential conduits for the spread of disease, but they are sometimes the cause of the problem. *Legionella* microorganisms, which enter and grow in HVAC systems, are the most notorious example. Less dangerous but far more typical is the growth of mold, the most common form of allergen. In the dark, moist environment of an HVAC system, mold spores can proliferate year-round. In allergic individuals, these spores initiate a chain of reactions starting with the release of histamines and inflammation of mucus membranes. These symptoms may lead to congestion, breathing difficulties, or even asthma and other complications. There are several categories of organisms that can grow and/or spread in modern air handling systems:

- Pathogens—viruses, bacteria, and fungi that cause a range of infectious diseases
- Allergens—bacteria and mold that cause allergic rhinitis, asthma, humidifier fever, and hypersensitivity pneumonitis
- Toxins—endotoxins and mycotoxins that cause a variety of toxic effects, irritation, and odors.

As HVAC systems move large amounts of outdoor and recirculated air through occupied buildings, they become the conduits by which these unhealthful organisms are spread throughout the spaces they serve.

In the ongoing quest for better indoor air quality (IAQ), experts have come to recognize that these biological contaminants in indoor air are major contributors to *sick building syndrome* and building-related illness. In fact, according to the World Health Organization, biological contaminants in buildings are believed to account for a substantial portion of absences from work and school as well as days where activity is impaired or restricted. As a comparison, in most cases, the cost of losses in productivity far exceeds the cost of operating and maintaining the HVAC system.

18.1.2 Clean Indoor Air Strategies

Biological contaminants are also among the most difficult to control. Though high-efficiency ASHRAE grade or HEPA filters are helpful, many systems do not lend themselves to filter upgrades without major changes. And since many microorganisms are typically less than 1 micron in size (with some viruses as small as 0.003 to 0.004 micron), even high-efficiency filtration may be inadequate. Another very important but overlooked issue is a condition that may occur when time-clock systems are turned off. Natural temperature differentials between the system and the space create a convection flow or "back-draft" effect that returns space contaminants back through the ductwork to the downstream side of the filters. When combined with system leaks, these conditions often compromise the filter's role.

Where biological contamination is known to exist, a common control strategy is duct cleaning, sometimes followed by a biocidal treatment. In subsequent swab sample testing, however, a biological activity of concern has often been demonstrated to return in as little as 3 months after cleaning and treatment. In cases where legionella microorganisms are present or suspected, acid washing (or other treatment) of fan-coils, drain pans, cooling towers, etc., is required—an expensive and often destructive procedure that shortens equipment life. Other methods for microorganism control tend to be impractical, potentially toxic, detrimental to equipment operation and efficiency, or simply too costly.

A *UVC light* system can be a significant control strategy to help reduce discomfort caused by microbiological reactions, *Legionella* microorganism growth and airborne dissemination, circulation of tuberculosis in air handling systems, and spread of cold and flu viruses. They kill harmful microorganisms without posing a risk to building occupants, maintenance personnel, mechanical equipment, and interior furnishings. If UVC emitters are to be used for general microbial control, *E. coli* is accepted as the standard target. However, if a specific organism is to be targeted, one must determine the right UVC dosage for that organism.

A kill rate of 90% (based on single-pass testing with conventional measurements) is accepted as the standard for general microbial control in IAQ situations. This rate will deliver nearly a log reduction in microbial contamination, achieving indoor air quality that is roughly equivalent to that of outdoor air. The light irradiance value—flux in microwatts per square centimeter—determines the number of photons available to strike a target microorganism. Reflectance within the cavity creates a global ricochet effect of generated photons, making a higher number of them available and bouncing them in all directions to penetrate every nook and cranny—in effect, potentially increasing the dosage received by any organism. Reflectance works literally to "unhide" these organisms for proper destruction.

The greater the irradiance, the higher the kill rate for that microorganism. For example, achieving a 99.99% kill versus a 90% kill mathematically requires four times the irradiance or number of UV sources or high reflectance. What's considered a good reflectant for visible light, however, is not always a good reflectant for the invisible light energy of UVC. For example, common glass totally attenuates UVC; therefore, a typical rear-surfaced glass mirror does not reflect UVC at all. Since a typical duct liner has little or no reflectance, practical solutions include coating surfaces with aluminum paint or lining them with aluminum foil or sheeting.

Air velocity, expressed in fpm, is a key factor in determining the time an organism spends within the physical cavity (sometimes referred to as "dwell time") and the amount of heat removed from the UVC lamp, which directly affects its output. Thus, the higher the velocity or lower the air temperature, the more UVC energy required to achieve desired performance. Humidity is another attenuator to UVC energy. If higher (greater than 60%) relative humidity exists, more UVC energy is needed to compensate for the absorption effect. Owing to variations in temperature, velocity, and other factors (as above), location has an impact on performance. For example, a UVC sterilization system upstream of the evaporator coil has different energy requirements than one located downstream of the coil.

Industrial Buildings and IAQ Control: In the industrial building environment where the quality of the air is a critical parameter, this refers only to the strict temperature and the moisture content. Moisture content is measured in terms of the mass of water vapor contained in a unit mass of dry air. The numerical values are quite small since at any particular temperature the air can only contain a finite amount of water vapor. The following are the main processes carried out in an industrial building air-conditioning plant that need to be controlled to maintain these air quality parameters in the plant. Refer to figure 4.2 Chapter 4

1. *Heating:* Heating causes the temperature to rise. The moisture content is not affected, however, and the relative humidity decreases with heating.

2. *Cooling:* Cooling causes the temperature to fall. Provided the temperature does not fall below the "dew point" of the mixture, there is no change in the moisture content and the relative humidity increases.

3. *Cooling below the dew point:* When the temperature of the mixture falls below the dew point, then the mixture can no longer hold the same mass of water vapor. The water vapor then condenses out of the mixture and appears on any cold surface which, in general is the source of the cooling. This degree of cooling is often used in air conditioning systems to remove water vapor from the mixture. Subsequent reheating if the mixture can result in the same original mixture temperature, but with a reduced relative humidity. Another way to achieve it conservatively, is through desiccant cooling as shown in Figure 18.1.

4. *Humidifying:* This process refers to the injection of water into the air to give an increased moisture content. Humidifying is usually associated with an attendant decrease in the mixture temperature. If the mixture is reheated, then it can be brought back to the same original temperature, but the relative humidity will be increased.

18.1.3   IAQ and Energy Conservation

In modern buildings, reduction of the fan-driving power is an indispensable part of any energy conservation program. Underventilation can create poor IAQ, while overventilation wastes energy if outside air must be heated or cooled. One solution is Demand Controlled Ventilation levels on the basis of measured quantities of CO_2 or other air quality parameters. There is sometimes valid overventilation—particularly in spaces with unpredictable swings in occupancy, such as auditoriums, cafeterias, theaters, retail stores, class rooms, and conference rooms.

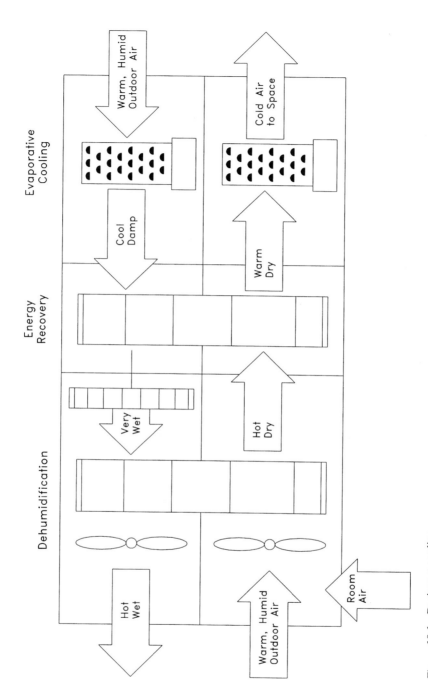

Evaporative Cooling

Energy Recovery

Dehumidification

Warm, Humid Outdoor Air

Cold Air to Space

Cool Damp

Warm Dry

Very Wet

Hot Dry

Hot Wet

Warm, Humid Outdoor Air

Room Air

284

Figure 18.1 Desiccant cooling.

Decentralized types of air conditioners are sometimes the best choice. These air cooling/reheating conditioners, which are concealed in the plenum, minimize the transport resistance. The use of a variable air volume (VAV) further enhances the reduction of the fan driving power. This allows separate outdoor and indoor air conditioners concurrently tempering the condition of outdoor air through energy recovery units (dehumidifying or heating and humidifying) and the indoor air conditioners cooling the sensible heat generated in a room.

The reason for this "division of labor" is because, as long as the two cooling systems are of a completely different nature, they should not be mixed. To be more specific, one system supplies chilled water of a relatively low temperature to the outdoor air conditioners, which then perform dehumidifying functions. The other system supplies chilled water of a relatively high temperature to the decentralized air conditioners, which then cool the sensible heat. The system as a whole improves the energy performance of the chiller, thereby contributing to energy conservation.

18.2   Water Resources, Quality, and Building Systems

One basic principle of the earth's hydrologic cycle is that water is a closed system; there is no new water input of any significance in the environment. (Some scientists may disagree with this principle, but on the whole, unless the Arctic ice caps melt, no significant increase in the earth's volume of water is likely to occur.) Two thirds of the earth's surface is covered by water. Of that, 96.8% is salt water in the oceans, and 3.2% is fresh water. Of the fresh water, 75% (2.4% of the earth's total supply) is frozen in ice and glaciers, another 24.4% (0.78% of the total) is in the form of ground water, and 0.6% (0.02% of the total) is in rivers, lakes, the atmosphere, and soil. The amount of water used by facilities and associated building systems is a very small portion of all the water on earth—less than 1%. To maintain the quality of this water economically, it is important to reclaim, recycle, and reuse our most precious natural resource surrounding our habitats.

18.2.1   Water Reclamation

Reclamation is defined as the conversion of naturally occurring nonusable water supplies into usable sources. Water reclamation generally falls into three broad categories: desalination, waste water reclamation, and surface and ground water reclamation. These reclamation activities generally take place on a municipal or regional level. However, in some remote industrial settings and owing to government regulations such as the Clean Water Act, many of these techniques and concepts are necessary to ensure a high-quality water supply. Building systems must be coordinated with reclamation systems so that there is an integrated and efficient approach to keep system cost down and provide an overall planned water system.

Given the vast quantities of salt water available, desalination provides rich opportunities to increase our supply of usable water. Common techniques for treatment include:

- *Multiple-effect distillation.* Heat transfer tubes are immersed in a tank filled with salt water, which is heated to the boiling point. It vaporizes and passes into a lower tank, where new incoming salt water cools it through a heat-exchange process, yielding a distillate of practically pure water.

- *Multiple-state flash.* Salt water is heated in a shell-and-tube heat exchanger, then passed into a second heat exchanger at lower pressure. The pressure change causes the salt water to "flash," or change back into a liquid form. The vapor is condensed, yielding a distillate of pure water.

- *Vapor-compression distillation.* In this process, salt water is heated in a heat exchanger and vaporized. Then energy is applied by a compressor, which takes the vapor and compresses it to a higher pressure and temperature. This, in turn, furnishes heat for vaporization of more salt water. The vapor is then condensed to yield distilled water.

- *Electrodialysis.* This desalination method is effective on waters with salinity below 28 ppt. The process pulls ions of dissolved salts out of the water by electrical forces, and these pass from the salt water compartment into adjacent compartments though permeable plastic membranes.

- *Reverse osmosis.* This method is effective for brackish water. Water is passed through membranes under high pressure, leaving the salts behind. On some locations, reverse osmosis desalination systems are used to reclaim high-quality fresh water from wells located below ground. Although these systems are expensive to install and operate, the fact that brackish well water is readily accessible within 50 to 75 feet below ground makes desalination more economical than drilling hundreds of feet to find fresh water, which may have to be treated in any event.

- *Other methods.* These include Freezing, solar evaporation, and ion exchange.

Additional treatment may be necessary depending on the water's intended use. For example, some industrial processes may require deionizing systems or other means to polish out specific contaminants that might be harmful to that process.

The second category for reclamation is waste water. Proper treatment of waste water is one way to help keep fresh water supplies from becoming contaminated, since the discharge of most of this water is returned back to the environment through irrigation systems, ponds, lakes rivers, oceans, and infiltration and percolation through the soil.

The third area for reclamation relates to surface and ground waters. Surface water is directly related to the amount of precipitation that can be expected annually. This water is not distributed evenly; much of it falls on mountainous areas and areas located near the tropics or equator.

Ground water is water stored at or above atmospheric pressure. (Water stored in the soil is under tension.) The amount of ground water stored is approximately 0.78% of the total fresh water on earth. Ground water is extremely important. In the United States alone it supplies approximately 25% of all the fresh water used for public supply, livestock, irrigation, industrial thermoelectric power, and other manufacturing uses, out of this 10% is used for building systems.

Treatment required for both surface and ground water varies, depending on the quality of the water and its contaminants. These contaminants include heavy metals, trichloroethylenes, nitrates, radonuclides, herbicides, pesticides, trihalomethanes and other organic materials such as dissolved solids, sodium, and chloride. The Safe Drinking Water Act also requires stricter regulations as it relates to contaminant levels. Refer to following EPA data on safe limit of contaminants in drinking water.

Pollutant	Maximum Contaminant Level (MCL) mg/liter or ppm
Arsenic	0.05
Bacteria	4/100 mL
Barium	1.00
Benzene (organic)	0.005
Cadmium	0.01
Carbon tetrachloride (organic)	0.005
Chloride	250.0
Coliform	<1/100 mL
Color (platinum–cobalt scale)	15 units
Copper	1.0
Chromium (hexavalent)	0.05
Cyanide	0.01
1,1 Dichloroethylene (organic)	0.007
1,2 Dichloroethylene (organic)	0.005
Endrin (organic)	0.0002
Fluoride	4.0
Foaming agents	0.5
Iron (>0.3 makes red water)	0.3
Lead	0.05

Pollutant	Maximum Contaminant Level (MCL) mg/liter or ppm
Lindate (organic)	0.004
Manganese (>0.1 forms brown-black stain)	0.05
Mercury	0.002
Methoxychlor (organic)	0.1
Nitrate	10.0
Odor (threshold odor)	3
p-Dichlorobenzene (organic)	0.075
pH	6.5–8.5
Selenium	0.01
Silver	0.05
Sulfate (SO_4)(>500 has a laxative effect)	250.0
Total Dissolved Solids	500.0
Toxaphene (organic)	0.005
1,1,1 Trichloroethane (organic)	0.2
Trichloroethylene (organic)	0.005
Trihalomethanes (organic)	0.1
Turbidity (silica scale)	1 to 5 TU
Vinyl choloride (organic)	0.002
Zinc	5.0
2,4–D (organic)	0.1
2,4,5–TP Silvex (organic)	0.01

Radionuclides:

Gross Alpha particle activity	15 pCi/L
Beta particle and photon radioactivity	4 mrem
Radium–226 and Radium–228	5 pCi/L

Exposures over safe limits can result in a variety of serious health problems ranging from liver and kidney damage, high cancer risk, nervous system disorders, skin discoloration, and hypertension.

Treatment options are extremely varied. Following is a summary of some of the most common methods:

- *Chlorination and ozonation:* Control pathogenic or nonpathogenic organisms
- *Activated carbon:* Removes dissolved organics, color, taste and odor-causing compounds.
- Flocculants: Make particles cling together for more efficient water filtration.
- *Mechanical filtration:* Removes suspended solids. These devices include gravity sand filters, slow-rate sand filters and high-rate sand filters, etc.

- *IR stripping/aeration:* Removes carbon dioxide, hydrogen sulfide and other taste- and odor-causing compounds, as well as volatile organic compounds.
- *Softening (ion exchange):* Removes hardness.
- *Chemical treatment:* Controls acidity.

In addition to basic treatment, many municipalities offer a reverse osmosis treatment system to remove trichlorethylene pesticides, nitrates, and radionuclide contaminants.

18.2.2   Water Recycling

Recycling refers to recovering water that would otherwise go into the waste stream and cleaning it for reuse. Ideally, the goal of the building systems engineer working in the commercial, industrial, and institutional arena should be to develop a closed-loop system by which waste water that is produced from human habitation could be recycled continuously with minimum introduction of fresh water into the cycle. Of course, to achieve this naturally, the population would have to be kept low enough relative to land density so that the natural biological processes can develop to allow a closed or semi-closed system to work in densely populated areas, which is not possible. Therefore, other ways to recycle water must be found. This includes recycling semigray, gray, and black water.

18.2.3   Water Reuse

Reuse refers to the reapplication of recycled water in other building systems or for landscaping, etc. Water reuse has been practiced since ancient times. Athenians reused waste water for crop irrigation. In the Far East, human waste has been used for thousands of years to fertilize ponds and produce aquatic animal and plant life. One large potential source of water for reuse in buildings is storm water. Many building sites have storm water retention, either in on-site ponds or underground storage tanks. This allows for a controlled discharge of storm water into sewers. It also allows storm sewers to be sized for average rainfall, thus reducing their installation cost. The incremental cost of recycling storm water is generally a fraction of the total cost of building a retention pond or storage tank. Thus, if such a storage facility is required, the cost of the additional pumps and piping for recycling should be evaluated. These storage basins also can be used for recycled gray and black water after on-site treatment. Commercial building water supply and drainage systems are shown in Figures 18.2 and 18.3. Conservation on these systems can be done as follows:

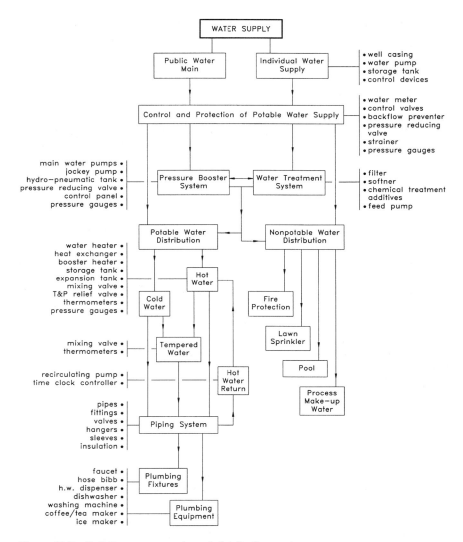

Figure 18.2 Building water supply and distribution system.

- *Flushing toilets and urinals.* This is an acceptable use for recycled waste water. Piping must be clearly marked "nonpotable" and a dye is added to the water for safety. This could potentially save 15% to 20% of building total water use per day. These types of systems should be considered for use in drought-prone areas.

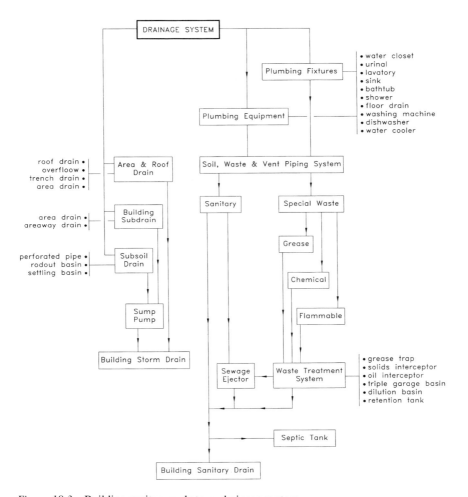

Figure 18.3   Building sanitary and storm drainage system.

• *Makeup water for cooling towers and mechanical cooling equipment.*
  Approximately 38% of the fresh water used in the United States is for
  these purposes. Therefore, saving 5% through recycling could result in
  a substantial savings in water bills per year.

Water conservation is not a plumbing issue alone. For example, reused water
may be used for fire sprinklers, and conservation certainly will have an impact
on electrical requirements for pumps, etc.

This type of integrated initiative offers the potential not only to save water but also to save energy costs associated with pumping water.

18.3   Power Quality and Intelligent Buildings

The concept of load and source compatibility is not new. The need to provide power with steady voltage and frequency has been recognized since the inception of the electric utility industry. However, the definition of "steady" has changed over the years, reflecting the greater susceptibility of increasingly sophisticated electronic equipment to the departure from "steady" conditions. Some of the early concerns were flicker of light bulbs due to voltage variations, and overheating of electromagnetic loads or interference of communication loads due to voltage waveform distortion. Recognition of these problems led to the development of voluntary standards that contributed significantly to reducing occurrences.

The advent of electronic power conversion has been widely applauded by users, but the drawbacks from the point of view of power quality have not always been recognized. The very advantages of solid-state devices that made possible modern switching power supplies, electronic ballast, invertor-rectifiers, high-frequency induction heating, and adjustable-speed drives also make these power converters into generators of harmonic currents and additional sources of line voltage drops. Thus, in addition to the disturbances generated by the normal operation of the familiar power delivery and load equipment, the disturbances resulting from the new electronic loads must be taken into consideration.

Power semiconductor devices that constitute the heart of modern power electronics have been undergoing dynamic evolution in recent years. Never before in the history of power semiconductor devices have we seen the emergence of so many exotic devices in such a short span of time. A power semiconductor device is indeed the most complex, delicate, and "fragile" element in a converter. The building system engineer needs to understand the device thoroughly for efficient, reliable, and cost-effective specification of converter. In addition, the cost of silicon-based power and control devices is continuously falling along with the improvement of performance, whereas the cost of passive system components, such as inductor, capacitor, transformer etc. is essentially constant and in some cases is gradually increasing, motivating integrated engineers to search for a "silicon solution" of passive components. Today's electronics industry is built on the integrated circuit—a tiny chip of silicon engraved with as many as a million microscopic electronic switches for controlling electric current. Because silicon's crystalline structure can be made to carry a nearly infinite array of electric pathways, it has become a vital electronic resource.

Only the purest silicon, however, can serve these purposes. Natural silicon consists of differently oriented crystals grouped into polycrystalline grains. Where these grains meet, they form irregular boundaries that can disrupt electricity flowing across them. To get around this problem, scientists devised a way to produce single-crystal silicon, a substance with a crystalline structure so uniform that it will accept any electrical pattern imposed on it.

The availability of high-power high-frequency devices at economical price will eventually permit mass application of motor drives, UPS systems, and active power line conditioners in building systems. HVAC engineers have always dreamed of using ideal switching devices in drives. Such devices should have large voltage and current ratings, zero conduction drop, zero leakage current in blocking condition, high temperature and radiation withstand capability, high mean time between failures (MTBF), and instant turn-on and turn-off characteristics. Of course, even with all these ideal features, the device should be available at an economical price. This dream will never materialize, but historically we have moved step by step in that direction.

It is important to understand what pure power looks like. Perfectly clean power will have a perfectly sinusoidal voltage of constant amplitude and frequency. Voltage amplitude will be adequate for the application, the voltage source will have no impedance, and the frequency will be 60 Hz (or 50 Hz in some foreign countries). The waveshape will be perfectly free of harmonics, noise, and transients. Of course, such a perfect source of power does not exist, even in a laboratory.

"Power quality" (strictly speaking, lack of quality) is a term often used today in describing an aspect of the electricity supply. A power quality problem is any occurrence manifested in failure or misoperation of electronic equipment. The newness of the term reflects the newness of the concern. Decades ago, when the traditional building system workhorse—the induction motor—was hit by a sag, it did not shut itself off, but produced a lower output until the sag was ended. Probably the most noticeable effect of a voltage reduction would be the dimming of the lights inside a building. But today, with the worldwide proliferation of sensitive power electronic equipment in building automation, these shortfalls in power quality can be very expensive in terms of building system shutdowns and equipment malfunctions.

18.3.1 Power Impurities

Let's have a closer look at some power impurities. The kinds of system events that contribute to the problem of power quality warrant closer examination, if only to better appreciate the remedies being proposed. So first, we will take a quick survey of everything from short interruptions to unbalance and flicker.

Voltage Variations: Voltage swells, which are brief increases in system voltage, can upset system electric/electronic controls and electric motor drives, including common adjustable-speed drives, which trip because of their built-in protective circuitry. Swells may also stress delicate computer components to the point of premature failure. Besides sags and swells, longer-lasting increases and decreases of voltage occur on occasion. An overvoltage has a less immediate effect than a swell, but it may shorten the life of building power system equipment and motors.

Undervoltages are sometimes due to the deliberate reduction of voltage by the utility in order to lessen the load during periods of peak demand. These planned undervoltages, often called brownouts, are the blight of those hot summer days when air conditioners are at full blast and utilities lack enough generation to keep them going.

Voltage disturbances even shorter than sags and swells are classified as *transients*. They fall into two basic classes: impulsive transients, attributable in many cases to lighting and building load switching, and oscillatory transients, usually due to utility capacitor bank switching. Utility capacitor banks are customarily switched into service early in the morning in anticipation of a higher power demand. With the exception of lightning, almost all transients are generated as the result of interaction between stored electrical energy in building power system inductances and capacitances.

When current is interrupted at peak current flow (for example), the building system inductive load (L) is left with considerable stored energy. When the flux collapses, this energy interacts with system capacitance (C), causing an LC circuit oscillation with a theoretical peak voltage of perhaps ten times the normal peak voltage. This type of interruption is known as "current chopping" and is a form of "current suppression." Current chopping refers to a vertical cut in the current wave and is caused by circuit breakers and switches interrupting light inductive loads such as the excitation current of an unloaded transformer. Current chopping is also common with high-speed circuit breakers. Current-limiting fuses, thyristor, or SCR switches used in building power system configurations commonly cause current suppression and generate large spikes or transients. The effect of transients on the computer can be errors due to the *dv/dt* (change in voltage/delta time) coupling through the stray and interwinding capacitances of the power supply.

Resonance voltage spikes are traveling waves on electrical circuits and follow all of the laws of transmission line theory. For this reason, it is extremely difficult to predict accurately their rise time, amplitude, or frequency of occurrence. At each change of impedance (such as a wire size change or splice) and at each transformer, a portion of the spike is reflected and a portion is refracted or passed through.

Another disturbance—and a very common one—is the *powerfail*. The powerfail is defined as the total removal of the input voltage for at least half a cycle per utility and manufacturer standards as shown in Figure 18.5. Powerfails can cause the floating heads of disk drives of building system computers to "crash down" on the disk, causing memory loss, unscheduled shutdown, or equipment damage. Some disk drives have heads that automatically retract upon loss of power, but the designer cannot assume that this is the case.

Harmonic Distortion: Harmonic currents are a result of building system equipment that require currents other than a sinusoid. See Figure 18.4 and word

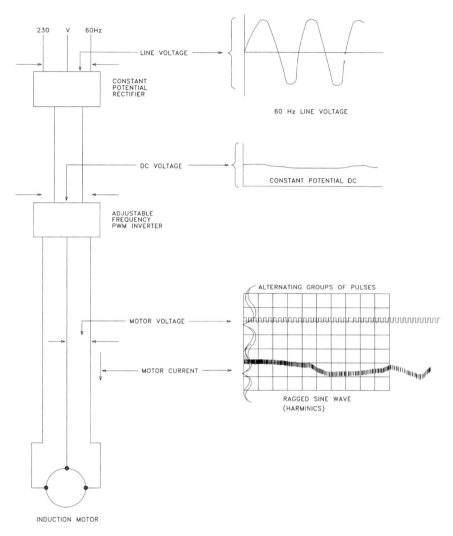

Figure 18.4 Simplified circuit, voltage, and current characteristics for a PWM adjustable frequency drive with fixed-pulse-rate system operating at reduced speed and generating harmonics.

harmonic on Figure 9.2 refer to components of building system that cause harmonic distortion. The amount of harmonic voltage distortion occurring on any building power distribution system will depend on the impedance versus frequency characteristic seen by the equipment and by the magnitude of the currents. The distortion factor can refer to either voltage or current. A more common

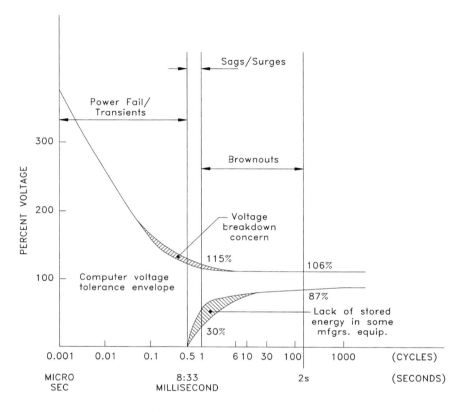

Figure 18.5 Typical power quality design goals. Note: This graph addresses only the magnitude of the voltage and its duration. It lacks information on rate of change in voltage (power) disturbance.

term that has come into use is *total harmonic distortion* (THD). To calculate THD the periodic wave is broken into its sinusoidal components, then a quantitative analysis of its parts is made. IEEE or IEC Standards make recommendations for limits within which current and voltage harmonics distortion should be kept. These standards are a system standard and not an equipment standard, and contain application information.

18.3.2 Effects on Building Systems

The major effect on building systems that have a power quality problem is increased heating due to iron and copper losses. It can give rise to high audible noise emission and can cause a phenomena called cogging (the refusal to start smoothly) or crawling (very high slip) in building system induction motors. Table 18.1 provides a summary of power problem along with possible causes and their effects.

A system resonance condition is the most important factor affecting system harmonic levels. Parallel resonance is a high impedance to the flow of harmonic current, while series resonance is a low impedance to the flow of harmonic current. In building systems where capacitor banks are used for voltage control and/or power factor improvement and the manner in which capacitors are connected can cause resonance conditions (both series and parallel) that can magnify harmonic current levels.

Parallel resonance occurs when the system current oscillates between the energy storage in the system inductance and the energy storage in the capacitance. The frequency at which parallel resonance occurs can be estimated by the following simple equation:

$$H_{\text{resonance}} = \frac{\text{short circuit MVA}}{\text{capacitor bank size in MVA}} = \frac{X_C}{X_L}$$

where H is the harmonic order, and X_C and X_L are capacitive and inductive reactances of building power system at the fundamental frequency.

Series resonance occurs as a result of the series combination of building power system capacitor banks and line or transformer inductances. Series resonance presents a low-impedance path to harmonic currents and tends to draw in, or "trap," any harmonic current to which it is tuned.

Building cables involved in system resonances may be subjected to voltage stress and corona which can lead to dielectric (insulation) failure. Cables that are subjected to "ordinary" levels of harmonic current are prone to parasitic heating. The flow of nonsinusoidal current in a conductor will cause additional heating over and above that expected for the RMS value of the waveform. The harmonic currents flowing through the resistance of the building power system represent heat as

$$P_{\text{loss}} = \sum_{h=1}^{h_{\text{max}}} I_{\text{harmonic}}^2 \cdot R_{\text{harmonic}}$$

R_{harmonic} for a given building power system can vary with applied harmonics because of skin effect, proximity effect, stray currents, eddy currents, etc. These vary as a function of frequency as well as conductor size and spacing. As a result of these, the effective alternating-current resistance (R_{AC}) is raised even higher, increasing the $I^2 R_{AC}$ loss.

Table 18.1   Summary of Power Problems—Cause and Effects

Power Problems	Possible Causes	Effects
High-voltage spikes and surges	Lightning Utility grid switching Heavy building system equipment	Equipment failures System lock-up Data loss
Low-voltage electrical noise	Electronic equipment Switching devices Motorized equipment Improper building power system grounding Power system protective devices, contractors and relays Copiers	Building system data corruption Erroneous command functions Variations in system timing signals Changes in building system processing states Loss of bilding systems synchronization Control instability Mal-activation of protective devices
Harmonics	Building system controllers Switch made power supplies Uninterruptible power supplies Building system motor drives Electronic lighting ballasts	High power system neutral currents Overhead distribution and power equipments, e.g., transformers, panelboards, neutral conductors.
Voltage fluctuations	Overburdened power distribution networks Power system faults Planned and unplanned brownouts Unstable generators	System lock-up Motor overheating System shutdown Bulbs burn-out Data corruption and loss Reduced performance Loss of control
Power outages and interruptions	Blackouts Backup generator start-up Backup battery system discharges	System crash System lock-up Lost data Loss of control Lost communication and system link Complete shutdown

18.3.3   Pure Power Solutions

The pure power need of intelligent buildings is analogous to its pure water need. Let's look into how water is purified. The purification of drinking water proceeds in several stages. The first step is sedimentation, in which large particles suspended in the water settle to the bottom. The second is filtration, in which suspended solids and harmful bacteria are strained out. In the third stage, chlorine, a powerful disinfectant, is added to the water to kill the remaining microorganisms. Unfortunately, chlorine can give water a bad taste and in large doses can even cause serious health problems.

Similarly impure power is purified in steps based on level and type of impurity and large doses of certain treatment type can cause serious power stability problems. If one provides a "cure for which there was no disease" the remedy can sometime cause disease. The most effective and least expensive way to remove building power system transient overvoltages is through surgery—simply cutting them off or clipping them. The problem is that of all the power conditioning techniques in use, this is probably the least understood and the most misapplied. Table 18.2 summarizes various types of power conditioners and their applications.

The most common lightning surge suppressor is the lightning arrester. Since the lightning arrestor responds slowly, the fast transients often passes through. Probably the most common low-voltage fast transient suppressor is the metal oxide varistor, which is fast enough to clamp or clip off most system transients. One of the disadvantages of system transient suppressors is that they pump transients to ground, thus shifting a line power quality problem to a building ground potential problem. Because of this, surge suppressors and lightning arrestors should always be separated from the building critical equipment ground by an isolation transformer. The shielded isolation transformer is often used in tandem with voltage regulators, transient suppressors, and other devices and power conditioners because it offers excellent power dirt rejection and a clean ground.

Most computers used in building systems do not need external voltage regulation. Those that do are generally served best by ferroresonant regulators which are extremely fast and do not "fight" with the computer regulator, or by very slow regulators that likewise avoid fighting with the computer. Where a high level of power quality and total isolation from sags, dips, surges, and transients is required, motor generator (MG) sets and uninterruptible power systems (UPS) are used. Since MG sets and UPS actually reconstruct the power sinewave by converting the system ac input voltage to another form of energy and then regenerate an ac voltage, these units are not pure power conditioners. UPS often require power conditioners themselves to prevent damage to their solid-state components.

Table 18.2 Summary of Performance Features for Various Types of Power Conditioning Equipment

Power Quality Condition		Transient Voltage Surge Suppressor	EMI/RFI Filter	Isolation Transformer	Voltage Regulator (Electronic)	Voltage Regulator (Ferroresonant)	Motor Generator	Standby Power System	Uninterruptible Power Supply	Standby Engine Generator
Transient voltage surge	Common mode	D		Y	D	Y	Y	D	D	
	Normal mode	D			D	Y	Y	D	Y	
Noise	Common mode		D	Y	D	Y	Y	D	D	
	Normal mode		D	D	D	Y	Y	D	Y	
Notches				D		Y	Y		Y	
Voltage distortion						D	Y		D	
Sag					D	D	D	D	Y	
Swell					D	Y	Y	D	Y	
Undervoltage					Y	Y	Y	D	D	
Overvoltage					Y	Y	Y	D	D	
Momentary interruption							D	Y	Y	
Long-term interruption										Y
Frequency variation								D	Y	D

Y (Yes)—It is reasonable to expect that the indicated condition will be corrected by the indicated power conditioning technology.

D (Doubtful)—There is a significant variation in power conditioning product performance. The indicated condition may or may not be fully correctable by the indicated technology.

Table 18.3   Matching Sensitive Load and Power Source Requirements with Expected Environments

A. Voltage

Voltage parameter affecting loads	Typical Range of Power Sources	Typical Immunity of Electronic Loads		Units Affected and Comments
		Normal Range	Pure Power Range	
Over and under-voltage	+6%, −13.3%	+10%, −15%	±5%	Power supplies, capacitors, motors; components overheating and data upset
Swells/sags	+10%, −15%	+20%, −30%	±5%	Same as above
Transients, impulsive and oscillatory, power lines	Varies: 100–6000 V	Varies: 500–1500 V	Varies: 200–500 V	Dielectric breakdown, voltage over stress; component failure and data upset
Transients, impulsive and oscillatory, signal lines	Varies: 100–6000 V	Varies: 50–300 V	Varies: 15–50 V	Same as above
ESD	<45 kV 1000–1500 V	Varies widely: 200–500 V	Varies widely: 15–50 V	Signal circuits; dielectric breakdown, voltage over stress, component, failure, data upset; rapid changes in signal reference voltage
RFI/EMI (conducted normal and common mode)	10 V up to 200 kHz less at higher frequencies	Varies widely 3 V typical	Varies widely 0.3 V typical	Signal circuits; data upset, rapid changes in signal reference voltage
RFI/EMI (radiated)	<50 kV/m, <200 kHz <1.5 kVm, >200 kHz	Varies widely w/shielding	Varies widely w/shielding	Same as above
Voltage distortion (from sine wave)	5–50% THD	5–10%	3–5%	Voltage regulators, signal circuits, capacitor filters, capacitor banks; overheating; undercharging
Phase imbalance	2–10%	5%		
			3%	Polyphase rectifers, motors; overheating

Table 18.3 (*continued*)

B. Current

Current Parameter Affecting Sources	Typical Range of Load Current	Typical Susceptibility of Power Sources		Units Affected and Comments
		Normal Range	Pure Power Range	
Power factor	0.85–0.6 lagging	0.8 lagging	<0.6 lagging or 0.9 leading	Power source de-rating or greater capacity source with reduced over-all efficiency
Crest factor	1.4–2.5	1.0–2.5	>2.5	1.414 normal; im-pact function of impedances at third and higher harmon-ics (3–6% Z); volt-age shape distor-tion
Current distortion	0–10% total RMS	5–10% total — 5% largest	5% max total 3% largest	Regulators, power circuits; overheat-ing
dc current	Negligible to 5% or more	<1%	As low as 0.5%	Half-wave rectifier loads can saturate some power sources; can trip circuit breakers
Ground current	0–10 A RMS + noise and surge currents	>0.5 A	<0.1 A	Can trip GFI de-vices, violate code, cause rapid signal reference voltage changes

C. Frequency

Frequency Parameter Affecting Loads	Typical Range of Power Sources	Typical Immunity of Electronic Loads		Units Affected and Comments
		Normal Range	Pure Power Range	
Line Frequency	±1%	±1%	±0.5%	Zero-crossing counters.
Rate of frequency Change	1.5 Hz/s	1.5 Hz/s	0.3 Hz/s	Phase synchroniza-tion circuits

18.3.4 Power Quality and Load Synchronization

Right now, providing power quality to specific data on utility systems is still in its infancy. Most probably, its course over the next few years will be directed by the economics of problems related to building system power quality and their solutions.

The regulated computer power supply has an input window that will vary from one unit to the next. The power supply will operate properly if the input voltage remains within the window. If power supply data for building computers are not available, the building system design engineer must insist on receiving data.

Table 18.3 gives indices of source power quality requirement in relation to load.

19

Integrated Building Systems Commissioning

The term commissioning entered the building system lexicon owing to the problems that owners increasingly encountered with new systems. As interest in commissioning grew, the focus remained mostly on activities that occur after systems have been installed and placed in operation. Commissioning of building systems literally means "simulating adverse conditions, performing and documenting test results and verifying the performance of systems so that systems interoperate in conformity with the design intent."

19.1   Introduction

Application of building automation systems in the commissioning process should be considered in every type of building where automation controls are installed. Improper calibration, installation, and unforseen interactions within applications can be detected during the commissioning phase if the automation systems are utilized to their full potential. Getting a building commissioned requires the complete cooperation between all involved parties, and any advantages that can bring an increase in efficiency to this process should not be overlooked.

Building system designers should include sections in their specifications that directly address the commissioning process. Specific response charts should be made to indicate how the automation system will be commissioned and how automation systems can assist in providing for total building acceptance. These kinds of specifications should include not only a list of what is to be commissioned, but also procedures for commissioning and the supporting documentation required. Before the first software program is written or installed into a controller, a commissioning plan should already be under development. Figure 19.1A and B are a recommended work sequence for commissioning authority selection and testing sequence.

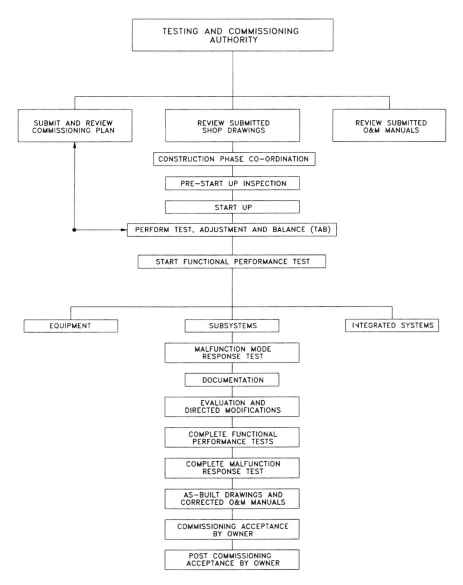

Figure 19.1(a)   Flow chart of work sequence for testing and commissioning authority.

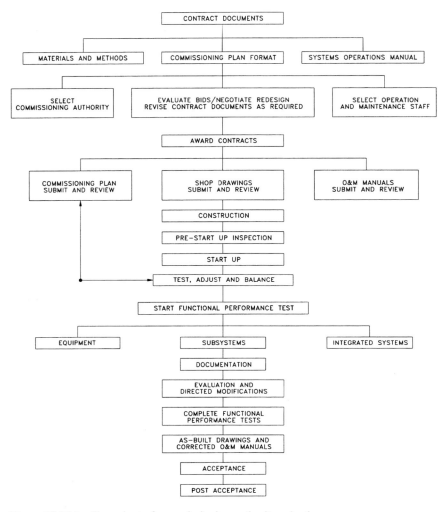

Figure 19.1(b) Flow chart of commissioning authority selection process.

If building system engineers do not give equal weight to the commissioning process as they are performing the design, the system effectiveness will already start slipping away. Successful commissioning hinges on successful design, including arrangements and layouts that permit proper installation with adequate clearance and access for operation, maintenance, and service; the sensors and readouts necessary to monitor the system and determine if it is doing what it should; and clear sequences of operation with measurable performance criteria.

Successful designs not only work in theory, but they also incorporate the features necessary to demonstrate proper performance during commissioning and daily operation. Each project requires several stages of building commissioning. The first step for all contractors is the visual inspection of the installation. Each piece of equipment (mechanical and electrical) must be verified per the job specification. Is it the right type? Has it been installed and calibrated correctly? Is standalone operation functional? For example, does a fan motor actually start the fan? Each contractor is typically responsible for the equipment he or she has installed.

19.2   Integrated Commissioning Design Needs

Designing for integrated commissioning involves most of the same insights as designing for a solo system. It needs proper access to equipment. If requirements are not clear from the engineering catalog, request a copy of the installation and operating manual. Understanding how the contractor will install the equipment and what parts require service help the engineer provide a design that accommodates those needs.

Design for integrated commissioning considers the performance implications of air intake locations (for ventilation or equipment operation) relative to exhaust discharges and other sources of contaminants. Commissioning goes beyond the plain language of the code to identify the underlying purpose of those requirements.

Overrides and Shutdowns: Part of the commissioning process often involves shutting down or isolating various equipment to simulate failures, alarm conditions, or seasonal changeover. Provide the means to accommodate it without disrupting other operations.

Think through to provide proper M/E integration. For example, if the HVAC system includes a free cooling heat exchanger, shutting down power to the chiller also kills power to the condenser water pump, thereby making free cooling unavailable. Yet the free cooling system was installed to make cooling available during chiller maintenance. Integrated commissioning looks for this type of conflict and avoids it whenever possible.

Of course, design for integrated commissioning also includes providing the isolation valves, switches, interlocking relays, and dampers that will permit the desired manual overrides and shutdowns without disrupting systems or equipment that need not be affected. For example, commissioning a remotely operated shunt trip devices on the main circuit breakers in an occupied building. The purpose of this device is to allow the fire department to shut off the power without entering electric rooms that might be on fire. Obviously, shutting off all electric power in an occupied building to test this function would be disruptive, possibly creating more danger to people than the shunt trip system was intended to prevent. Even if the test is run outside of regular hours, interrupting power would wreak havoc with digital clocks and other electronic devices throughout the building. The integrated solution is to provide a test mode with LED display that lights a light bulb instead of shutting off all power in the building. Design for integrated commissioning includes thinking ahead to how to test various systems and providing means to simulate operating or alarm conditions or activate "dummy" outputs.

A big part of commissioning involves measuring and evaluating data obtained from various sensors. The commissioning team will bring some instruments of their own, but they will also rely on devices permanently installed as part of the system. Whether those devices are pressure gauges, current transformers, potential transformers, thermometers, or building automation and control system sensors, they get installed only if the design engineer calls for them. Therefore to design for integrated commissioning, the design engineer must consider the types of temperature, pressure, current, potential, and other sensing devices that will be needed for commissioning and daily system operation and include them on the plans and specifications.

As a general rule, the commissioning team will want to observe at every piece of equipment or device where those parameters change. Some examples are electrical parameters on motors for compressors, chillers, pumps, fans, and corresponding pressure, temperature, and flow parameters on pumped, boosted, or compressed media.

The commissioning team will go through the system at startup and perhaps again at an opposite season to verify system performance. Integrated design that considers commissioning needs also benefits operating personnel. The cost to provide the sensors needed for commissioning is not a one-time expense; it is an investment in tools for conscientious system operation.

Often it is not cost effective or even practical to include thermometers and pressure gauges at every coil or baseboard radiation run. At these smaller loads, pressure/temperature taps (P/T or "Pete's" plugs) might be a reasonable compromise. But a P/T plug on a pipe between a duct and a wall where there is no room to put a ladder or read the gauge might just as well not be there. The sensor location is also critical for commissioning. For example, if a temperature sensor is located in a stagnant or stratified air pocket, it can give inaccurate input into a DDC control loop and cause unexpected output. The same can be said about an improperly calibrated sensor.

Some locations need flow meters that continuously read and display instantaneous flow, either directly or through a building automation system. Obviously, these features add cost over simple head-type devices (pitot tube, orifice, etc.) that require a differential pressure meter to read. As a general rule, the larger the piece of equipment, the more energy it uses; or the more important it is to the overall system, the more it needs a flow meter. The integrated commissioning team verifies the flow by meter, against current drawn by motor and also relates it to pressure differential indicated by gauges on the equipment.

Commissioning plans for VAV systems include evaluating the system static pressure sensor performance of this control. The design engineer often treats it by a rule of thumb such as "two-thirds to three-fourths down the duct," possibly making vague reference to a similar comment on ASHRAE Handbook Systems. The more likely location for the system static pressure sensor is near the static pressure or hydraulic midpoint of the system. Depending on how the engineer laid out and sized the ducts, that point could be much less than two-thirds of the physical distance from the fan to the last zone on that branch or it could also be much further away.

The same analysis applies to variable speed pumping. The engineer who designs for commissioning calculates the pressure differential required at the hydraulically most remote point and selects a suitable sensor location. The location must be one where calculations predict that a constant pressure differential at the sensor will always provide the hydraulically most remote point with at least the minimum pressure differential it needs for design flow.

19.3   Intelligent Building Automation Commissioning

As microelectronic controls continue to proliferate in all market segments of commercial buildings, even relatively small projects now can justify computer-based energy management systems. Design engineers must decide which functions to perform by traditional hard-wiring and which to perform by software logic, to fully utilize their potential in the process of building commissioning. Using automation controls with careful coordination of owner needs, electrical and mechanical disciplines and contractors installation practices can provide relief in this area.

For building system automation, the first stage of commissioning involves testing, calibrating, and verifying the sensors and actuators/operators installed on the job. Verifying wiring terminations; checking the placement of the sensors, controllers, operators, and actuators; calibrating sensor readings against known values; and stroking actuators should be done before automatic control loops are commissioned.

Verifying that building systems follow the integrated design sequences of operation and confirming satisfactory performance are important parts of integrated commissioning. To accomplish these goals effectively, the commissioning team needs a clear statement of the design sequences of operation. The sequences of operation should include measurable interoperating performance criteria so the commissioning team will know when the system achieves the design intent.

Starting at the coldest outdoor temperature and proceeding to the hottest, the design engineer describes what the system is supposed to do. Similarly starting at the fraction of a second power loss to momentary power loss and proceeding to extended power loss, the integrated design engineer describes what the backup system is supposed to do and how much time is involved to avoid a race between restart of original equipment and start of backup equipment. This avoids chaotic operation of bypass resulting from a false conclusion of failure.

The engineer also must think about how to describe different system responses. For example, saying that on control failure the HVAC system shall go to full cooling has no meaning. What if it is really a power failure, monitored as control failure and the system goes to full cooling? In this scenario, when the standby generator comes on line it trips on overload owing to all redundant equipment running on the generator under a false control signal. Does the engineer really want to override system operation this way and cause shutdown in spite of generators? This is where integrated commissioning can help tremendously.

With the wide availability of computer-based energy management systems, both design engineers and control contractors tend to ignore the details of how to implement the sequences during the design and submittal phase of a project. They brush off questions about coordination and sequencing of devices with the assurance that everything can be handled by software. Integrated commissioning questions then get resolved haphazardly in the field, possibly by people with no understanding about the integrated design intent.

As the name implies, many sequences of operation call for sequencing various devices so one device does not begin its action until the previous device has completed its action. With pneumatic controls, implementing this sequence is a relatively simple matter of selecting appropriate spring ranges. In theory, it should be possible to sequence three or four pneumatic devices with several electric devices using pilot positioners or ratio relays to modify the effective control range of the device. As a practical matter, sequencing more than two pneumatic devices or adding more than one electric device is usually problematic. All pneumatic devices have some hysteresis, so a nominal 5 psi spring range is probably somewhere between 4 psi and 6 psi, with good maintenance. All PE switches need some measurable differential between on and off. If a sequence of operations using pneumatic controls has more than two stages of pneumatic control, the engineer might be well advised to find a way to separate the control into independent loops.

The second stage in automation commissioning involves verifying and testing local control of the installed equipment, based on a specified sequence of operation. Local control covers those parts of the sequence of operation that are self contained in one controller. For example, the control of a VAV box is normally contained in one microcontroller, and an air handling unit is normally handled by a single, larger unit controller.

Larger unit controllers have multiple control loops running at the same time. Verifying the correct operation of an application requires breaking the application into smaller pieces that can be isolated and tuned, then tuning all the loops together. By this procedure, it is much easier to identify problems and correct them quickly. Once this has been done for all controller outputs, the interactions between isolated pieces must be tested. This step usually involves adjusting loop cycle times, tuning PID algorithms, and checking the output against the desired response from the isolated application.

The final stage of automation commissioning checks system-wide applications and interactions. System-wide operation requires the interactions of many different controllers and pieces of equipment. Typical applications include power demand strategies, power failure coordination, energy optimization and smoke removal and fire alarm sequences, and system-wide monitoring functions. Proper transmission and reception of the data must be ensured; response lag times must be accounted for; and failure conditions must be checked in multiple locations.

It is clear to see that each stage of commissioning relies heavily on the previous one. However, there is no guarantee that a single pass through the commissioning procedure will result in a working, stable building. Fine-tuning of a building automation system is an interactive process that requires the ability to isolate variables and identify trends over a period of time.

19.4   Personal Computers as Commissioning Diagnostic Tools

With the increased power of personal computers (PCs) and their cost decreases, building automation system data gathering functions have migrated to an affordable PC platform. These central data gathering stations not only monitor physical data, but also indicate whether or not a particular application is behaving correctly. In addition, they supply diagnostic and troubleshooting information, and document test results.

One of the best uses of PCs is to store cyclic readings of physical data to analyze long-term operation of a specific application or group of applications. Storing this type of information is referred to as trend logging. Identifying the trends in data, recognizing patterns, and isolating variables are the quickest ways to diagnose and troubleshoot operational stability.

For example, to test that a group of fan-powered VAV boxes in a specific area has been properly commissioned and is operational, the building operator can trend and log usually to a printer or a diskette, the physical data related to those VAV boxes. These data might include space temperature, temperature setpoint, fan motor current draw, damper position, and air flow. After a reasonable time span, it can then be analyzed to check for operational stability.

In applications that have a relatively slow reaction to change such as room relative humidities, the sampling time for data collection can be on intervals ranging from few minutes to an hour. In such applications, where the request for data to fill in the trend log is relatively infrequent, multiple trend logs are allowed to operate concurrently. This increases the efficiency of the commissioning process by allowing an operator to check more than one system at the same time.

For those applications that have much faster response times and require sampling rates of 30 seconds to two minutes, dynamic trending can be used. Dynamic trending differs from trend logging in that the data usually displays in a separate window on the operator's PC screen as a dynamic plot. Also, because the request for data is much more frequent, the number of concurrent dynamic trends that can be run at one time is limited.

Where more than one variable can command an output, the command trace feature can show the operator of a building system what action commanded an output and when this action occurred. For example, an air handling unit supply fan might have the following command variables: operator direct command, fire alarm overrides, optimum start/stop program, distributed power demand, and a time schedule. Knowing which one of these programs actually commanded the fan and when it was commanded helps determine if its operation is correct or not.

Monitoring the long-term financial performance of building system operation is of great interest. PC-based programs that monitor energy consumption, in combination with system trend logs, can give building owners a more global picture of how well their buildings are being run. Optimization of control strategies to minimize energy use can then be developed and tested to provide the best control at the lowest possible cost.

As PC-based central stations move toward running multitasking operating systems and support high-speed communication links to the controllers, application packages and expert systems that can receive and analyze large volumes of data are the next logical step. Not only could these types of applications constantly monitor the performance of the entire building automation system, but they could also predict problem areas and automatically take the necessary corrective action. For example, a program that monitors and analyzes all air flows and temperatures at VAV boxes will not only automatically fine tune the temperature reset to optimize energy consumption but will also warn the user when a sensor is out of calibration and provide control stability of the outputs based on a self-learning algorithm. These types of functions would provide increased diagnostic and system operation stability capabilities.

20

Building System Microeconomic Analysis

20.1 Introduction

Analysis of overall owning and operating costs and comparisons of alternatives requires an understanding of the energy units, electrical/mechanical losses, utility billing structures, cost of lost opportunities, inflation, interest, discounts, depreciation, and their relationship to the time value of money. This process, from basic system unit integration to losses collection and proceeding to economic analysis with full understanding of utility billing structures, is called microeconomic analysis. Owning and operating and maintaining over a period of time is called life-cycle cost analysis.

20.2 Integration of Energy Units

Over the past two decades, the large building systems have made dramatic progress in cutting energy consumption. However, documentation of the results, especially on an individual facility basis, has often been invalid and even misleading. That is because the techniques used for calculating energy saving fail to distinguish between the different forms of energy.

Presented here are calculation methods for converting power, steam (at different temperatures and pressures), and various types of fuels to a single common basis: methane. Btu subscripts are used to assist in following the logic. When a building owner purchases all his electricity from a utility and makes all his steam in boilers using only pure methane as the fuel, energy accounting is relatively straightforward. The calculations become more complex, however, when large complex building systems cogenerate steam and electricity, and use a mixture of industrial fuels. Understanding the relative values of fuel, electricity, and different levels of steam is especially important when justifying and evaluating cogeneration plants as part of building system.

The techniques described here are applicable to a single large facility or a site that includes many buildings of an entire company. A technically sound methodology allows valid comparisons among building systems that use different levels of steam, different forms of energy, and various types of energy-conversion devices. The energy consumption of steam or gas turbine drives, for example, can be compared with that of electric motors.

The procedures described are not intended to serve as a model for setting steam and electricity prices, although they may be applied to allocating fuel costs. Many factors not directly related to energy consumption enter into setting the price for a pound of steam or a kilowatt-hour of electricity. Operating and maintenance costs, for example, together with distribution costs and capital depreciation, are normally added to the fuel cost associated with generating electricity. For consistency and clarity, Btu and lb are used. In countries where different systems of units are employed, the procedures presented here will need only simple unit conversions.

20.2.1 Common Base Conversions

In calculating the energy consumed by a system, one usually needs to add the energy content (Btu) of fuel of various compositions, the energy content (Btu) of steam at various temperatures and pressures, and the energy content (Btu) of electricity that is purchased or generated at the site. These are different types of Btu. To avoid the problem of adding "apples and oranges," one must convert these individual Btu to a common base.

Typical natural gas used in the buildings include methane (70–98%), ethane (1–14%), propane (0–4%), butane (0–2%), pentane (0–0.5%), hexane (0–2%), carbon dioxide (0–2%), oxygen (0–1.2%), and nitrogen (0.4–17%). The composition of natural gas depends on its geographical source. Local gas utilities are the best source of gas composition. For a common base, the higher heating value (HHV) of methane is commonly used because methane is closest to the natural gas value, it is a widely used fuel, and its properties do not vary with geographic location and they are well documented. The reason for choosing HHV, instead of lower heating value (LHV), is that methane is normally purchased on HHV basis in the United States. In much of the rest of the world, however, LHV is the standard. For purpose of energy accounting, standardize on one or the other.

The two types of heats of combustion for fuels sometimes cause confusion. The difference can be traced back to whether H_2O produced during combustion is referenced to the liquid or vapor state. For methane:

$$CH_4 + 2\,O_2 \rightarrow CO_2 + 2\,H_2O$$

HHV = 1013 Btu/std. ft³ of methane (with respect to H_2O in liquid state at 60 °F and 1 atm). LHV = 913 Btu/std. ft³ of methane (with respect to H_2O in vapor state at 60 °F and 1 atm) The difference = 100 Btu/std. ft³ of methane (needed to vaporize water at 60°F and 1 atm).

When carbon monoxide is used as a fuel, no water is produced. Therefore, the HHV and LHV of CO are the same.

$$CO + 1/2\ O_2 \rightarrow CO_2$$

HHV = 322 Btu/std. ft³ of CO; LHV = 322 Btu/std. ft³ of CO.

The methods described below show how to convert nonmethane fuel, steam, and electricity to their methane equivalents. For consistency, the HHV of methane is used throughout.

A. From Btu of Fuel to Btu_{meth}

In practice, one is interested in the energy value corresponding to the LHVs from fuels because H_2O formed during combustion goes up the stack in the vapor state (along with the heat of H_2O vaporization). Therefore one could argue that it makes more sense to determine methane equivalents of fuels on the basis of the LHV of methane. For the reasons given earlier, however, the HHV of methane has been chosen here. Therefore, it is necessary to convert the LHV of various fuels to their CH_4 equivalents relative to the HHV of methane.

An electrically similar approach is used to calculate available short-circuit currents when the distribution system contains several voltage levels. Electrical quantities expressed as per unit on a defined base can be combined directly, regardless of how many voltage levels exist from source to fault. For an overall system approach this is easily accomplished by multiplying the LHV of a fuel by the ratio of the HHV of CH_4 to the LHV of CH_4. The resulting CH_4 equivalent relative to the HHV of methane is abbreviated as HHV_{meth}.

For propane:

$$HHV_{meth} = \frac{(\text{HHV of } CH_4)}{(\text{LHV of } CH_4)} \times (\text{LHV of propane})$$

$$= (1013/913 \times 2385 = 1.11 \times 2385$$

$$= 2647\ Btu_H/\text{std. ft}^3$$

For CO:

$$HHV_{meth} = \frac{(\text{HHV of } CH_4)}{(\text{LHV of } CH_4)} \times (\text{LHV of CO})$$

$$= 1.11 \times 322 = 357\ Btu_H/\text{std. ft}^3$$

Table 20.1   Types of Heating Values

Fuel	LHV (Btu$_H$/std. ft^3)	HHV (Btu$_H$/std. ft^3)	HHV$_{meth}$ (Btu$_{meth}$/std. ft^3)	HHV Difference (%)
Methane	913	1013	1013	0%
CO	322	322	357	+10.9%
Hydrogen	275	325	305	−6.2%
Ethane	1641	1792	1821	+1.6%
Propane	2385	2592	2647	+2.1%
Butane	3113	3373	3455	+2.4%

Comparison of heating values for various fuels shows the difference between lower and higher heating values and the methane-equivalent HHVs.

The HHV, LHV, and HHV$_{meth}$ for common gaseous fuels are given in Table 20.1. We have used Btu$_H$ to indicate that these re-enthalpy (heat) Btus. Note that the HHV$_{meth}$ of CO is higher than either its LHV or HHV (357 versus 322 Btu$_H$/std ft^3). Most fuels are more valuable (in terms of HHV$_{meth}$) than their HHVs would suggest. This is because most fuels, when burned, produce less H$_2$O during combustion.

Propane or butane and air mixtures are used in place of natural gas in small communities.

B. From kWh of Electricity to Btu$_{meth}$

The appropriate factor for converting electricity to methane equivalents depends on the source of electricity. If electricity is purchased from a utility, the logical thing to do is to use the efficiency of the power plant supplying the electricity. Because of the national power grid in the United States, however, the exact source of electricity may be difficult to determine. The average power-plant efficiency in the United States is about 34% (HHV basis), and is normally expressed as a "heat rate" of 10,000 Btu$_{meth}$/kWh. Unless there is a specific reason for using a different value, standardizing on 10,000 Btu$_{meth}$/kWh for purchased power in the United States is recommended.

When power is generated at a plant site, it is normally cogenerated with steam. In this case, the efficiency of the cogeneration plant should be used for the portion of electricity coming from the cogeneration plant. Any additional power that is purchased should be valued at 10,000 Btu$_{meth}$/kWh. If cogenerated power is sold to a local utility or to another operating company, the fuel attributable to power sales should be deducted from the site's energy consumption. The quantity of fuel is calculated by multiplying the power sales (kWh) by the heat rate of the cogeneration plant (Btu$_{meth}$/kWh).

If a site imports and exports power at various times during the year, imported power is valued at 10,000 Btu$_{meth}$/kWh. Exported power is valued at the actual efficiency of the cogeneration plant. Over the course of a year, the fuel attributable to power distributed within a plant site should be calculated using the weighted-average heat rates of purchased and cogenerated electricity.

As an example, suppose a site purchases 20 million kWh annually from the electrical grid (at 10,000 Btu_{meth}/kWh) and generates 100 million kWh in its own cogeneration plant where the heat rate is 9000 Btu_{meth}/kWh. During the year, 10 million kWh is sold to the local utility. The weighted-average heat rate for power distributed within the site is calculated as follows:

$$\text{Fuel for purchased power} = 20 \text{ million kWh} \times (10{,}000 \ Btu_{meth}/kWh)$$
$$= 200{,}000 \text{ million } Btu_{meth}$$

$$\text{Fuel for cogenerated power used} = 90 \text{ million kWh} \times (9000 \ Btu_{meth}/kWh)$$
$$= 810{,}000 \text{ million } Btu_{meth}$$

Total fuel attributable to site power use
$$= 200{,}000 \text{ million } Btu_{meth}$$
$$+ \ 810{,}000 \text{ million } Btu_{meth}$$
$$= 1{,}010{,}000 \text{ million } Btu_{meth}$$

$$\text{Total power distributed to site} = 20 \text{ million kWh} + 90 \text{ million kWh}$$
$$= 110 \text{ million kWh}$$

Weighted-average heat rate of power distributed to site
$$= 1{,}010{,}000 \text{ million } Btu_{meth}/110 \text{ million kWh}$$
$$= 9182 \ Btu_{meth}/kWh.$$

Note that we are not using the standard textbook conversion factor of 3413 Btu_H/kWh. This is a theoretical conversion factor. While it is possible to convert electricity into heat at close to 100% efficiency, we cannot convert heat (or, more specifically, the heat of combustion of a fuel) into electricity at 100% efficiency. Thus, no power plant operates at a heat rate of 3413 Btu_{meth}/kWh. The average United States power plant efficiency of 34% is calculated by dividing the theoretical conversion factor of 3413 Btu_H/kWh by the average power plant heat rate (typically 10,000 Btu_{meth}/kWh). Thus, $[(3413 \ Btu_H/kWh)/(10{,}000 \ Btu_{meth}/kWh)] \times 100 = 34\%$.

C. From lb of Steam to Btu_{meth}

If steam is generated in a boiler, the fuel used in the boiler—expressed as equivalent methane (in Btu_{meth})—is charged directly to the steam produced. If the steam pressure is reduced through a valve from, say, 400 to 30 psig (the same fuel equivalent as Btu_{meth}/lb of steam) is applicable to both levels of steam. No further calculations are necessary.

In a cogeneration unit, however, the situation is more complex. Here converting cogenerated steam to methane equivalents is a two-step process. First, the given quantity (in lb) of steam is converted to its equivalent kWh by using a thermodynamic function known as available work and applying an efficiency factor. Then the steam kWh is converted to Btu_{meth}, using the efficiency factor of the cogeneration facility. This two-step procedure is outlined below:

1. *From lb of steam to equivalent kWh*

The formula for calculating the maximum work available from a steam is given by:

$$W_{max} = \Delta H - T_0 \Delta S$$

or

$$W_{max} = (H - H_0) - T_0(S - S_0)$$

The enthalpy and entropy values for steam are readily available from steam tables. Selecting a heat-sink temperature is somewhat arbitrary. In principle, it is the lowest temperature available for cooling. In practice, however, this value changes daily. For consistency, a heat-sink temperature of 101 °F (560.67 °R) is recommended. (In cold climates, a lower heat-sink temperature may be selected.) This is analogous to the 2-in. Hg standard used for rating steam turbines.

The maximum available work from 1 lb of 1250-psig steam at 950 °F is calculated as follows:

$$W_{max} = (H - H_0) - T_0(S - S_0)$$

For 1250-psig steam at 950 °F:

$$H = 1468.0 \ Btu_H/lb$$

$$S = 1.60210 \ Btu_H/lb. \ °R$$

For saturated steam at 101 °F and 20 in. Hg:

$$H_0 = 1105.5 \ Btu_H/lb$$

$$S_0 = 1.97970 \ Btu_H/lb \ °R$$

$$T_0 = 560.67 \ °R$$

$$W_{max} = (1468.0 - 1105.0) - 560.67 (1.6021 - 1.9797)$$

$$W_{max} = 574.7 \ Btu_W/lb$$

To convert *maximum* available work to *actual* available work, multiply by 0.80. This gives a realistic amount of work that could be accomplished by the steam. (Modern steam turbines are capable of achieving 80% efficiency. If a facility has less efficient turbines, a lower percentage value may be more appropriate.)

$$W_{act} = 0.80 \times W_{max} = 459.8 \text{ BTU}_W/\text{lb}$$

To convert to kWh/lb, divide W_{act} (in Btu_W/lb) by 3413 Btu_W/kWh:

$$W_{act} = (459.8 \text{ Btu}_W/\text{lb})/(3413 \text{ Btu}_W/\text{kWh})\} = 0.1347 \text{ kWh/lb}$$

Note that the use of 3413-Btu_W/kWh is legitimate here because this conversion factor represents a theoretical maximum of 100% efficiency. It can also be expressed as 3413 Btu_H/kWh.

2. *Convert steam kWh to* Btu_{meth}.

To convert steam kWh to Btu_{meth}, multiply by the efficiency of the cogeneration facility. In the next section, we describe the technique for calculating the efficiency (heat rate) of a cogeneration plant. In the example calculation below, a heat rate of 10,300 Btu_{meth}/kWh is assumed.

$$\text{Btu}_{meth}/\text{lb} = 0.1347 \text{ kWh/lb} (10,300 \text{ Btu}_{meth}/\text{kWh}) = 1387.5 \text{ Btu}_{meth}/\text{lb}$$

Note that 1250-psig steam at 950 °F has three different Btu/lb values and that each has its own reference temperature:

$$1468.0 \text{ Btu}_H/\text{lb (above an arbitrary 32 °F base)}$$

$$574.7 \text{ Btu}_W/\text{lb (using a 101 °F heat-sink temperature)}$$

$$1,387.5 \text{ Btu}_{meth}/\text{lb (above a 60 °F base)}$$

Using subscript notation for these three different types of Btu helps avoid confusion.

Efficiency of a Cogeneration Plant: To assign appropriate Btu_{meth} values to steam and power produced by cogeneration, we must know the efficiency of the cogeneration plant. Using the available-work approach described above, this is a straightforward calculation. The idea is to convert various levels of steam production into equivalent kWh so that they can be added to the kWh of electrical output. To understand the procedure, see Figure 20.1 and the sample calculation in next section.

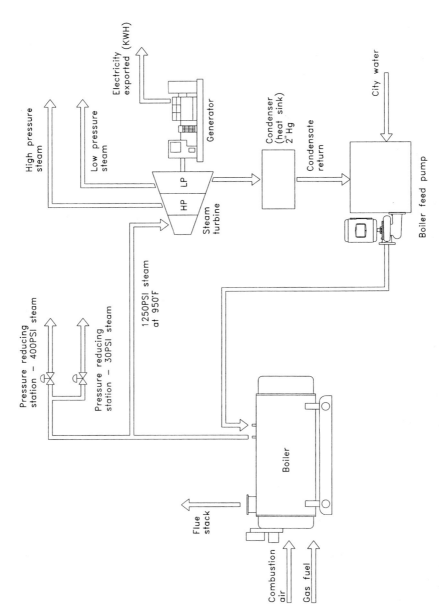

Figure 20.1 Congeneration system. A congeneration plant typically exports electricity and one or more levels of steam.

320

20.2.2   Sample Conversion Calculations

A. Suppose we have a cogeneration plant (Figure 20.1) with the following annual performance:

Fuel used = 26,280,000 million Btu_{meth}

Net electrical energy exported = 2190 million kWh

Steam exported = 1752 million lb of 400-psig steam at 650 °F, and 2628 million lb of 200-psig steam at 500 °F.

Step 1. Convert steam exported to equivalent kWh:

$$W_{max} = (H - H_0) - T_0(S - S_0)$$

For 400-psig steam at 650 °F:

$$H = 1333.9 \ Btu_H/lb$$

$$S = 1.61066 \ Btu_H/lb \ °R$$

For 101 °F saturated steam at 20 in. Hg:

$$H_0 = 1105.5 \ Btu_H/lb$$

$$S_0 = 1.97970 \ Btu_H/lb \ °R$$

$$T_0 = 560.67 \ °R$$

$$W_{max} = (1333.9 - 1105.0) - 560.67 \ (1.61066 - 1.97970)$$
$$= 435.8 \ Btu_W/lb$$

$$W_{act} = [(435.8 \ Btu_W/lb)/3413 \ Btu_W/kWh)] \ (0.80)$$
$$= 0.1022 \ kWh/lb$$

Total actual work = 0.1022 kWh/lb (1752 million lb)
$$= 179 \ million \ kWh$$

For 200-psig steam at 500 °F:

$$H = 1267.2 \ Btu_H/lb$$

$$S = 1.61476 \ Btu_H/lb \ °R$$

For 101 °F saturated steam at 2 in. Hg:

$$H_0 = 1105.5 \text{ Btu}_H/\text{lb}$$

$$S_0 = 1.97970 \text{ Btu}_H/\text{lb °R}$$

$$T_0 = 560.67 \text{ °R}$$

$$W_{max} = (1{,}267.2 - 1105.0) - 560.67 (1.61476 - 1.97970)$$

$$W_{act} = [(366.8 \text{ Btu}_W/\text{lb})/3{,}413 \text{ Btu}_W/\text{kWh})] (0.80)$$
$$= 0.0860 \text{ kWh/lb}$$

Total *actual work* = 0.0860 kWh/lb (2628 million lb) = 226 million kWh.

Step 2. Divide fuel use by sum of power kWh and steam kWh.

The total equivalent power output = exported power + 400 psig steam + 200 psig steam = 2.190 million kWh + 179 million kWh + 226 million kWh = 2595 million kWh.

The heat rate for the cogeneration plant = (26,280,000 million Btu_{meth})/(2595 million kWh) = 10,127 Btu_{meth}/kWh.

This heat rate can now be applied to steam and power used by an individual plant, and to electricity sales. If, for example, the cogeneration plant sells 5 million kWh to a local utility company, the fuel allocation would be: 5 million kWh × 10,127 Btu_{meth}/kWh = 50.635 million Btu_{meth}

B. Purchased Electricity and Steam Made in Boilers

If electricity is purchased from an outside utility and steam is produced in boilers, one can use a heat rate of 10,000 Btu_{meth}/kWh (the U.S. average) for electricity, and a typical boiler efficiency. In the following example, the methane equivalent of 400-psig stem is calculated using a boiler efficiency (HHV basis) of 70%. Condensate enters the boiler at 150 °F.

Step 1. Calculate the heat added to the steam in Btu_H.
For 400-psig steam at 650 °F:

$$H = 1333.9 \text{ Btu}_H/\text{lb}$$

For condensate at 150 °F:

$$H = 118.0 \text{ Btu}_H/\text{lb}$$

Heat added = 1333.9 Btu_H/lb − 118.0 Btu_H/lb = 1215.9 Btu_H/lb steam.

Step 2. Calculate the fuel required using a 70% thermally efficient boiler.

Fuel required = (1215.9 Btu_H/lb)/[(70 Btu_H)/(100 Btu_{meth})] = 1737.0 Btu_{meth}/lb of steam. Note that we can divide the amount of heat added to the steam (1215.9 Btu_H/lb of steam) by the furnace efficiency, which is expressed as Btu_H/Btu_{meth}.

If 200 and 30 psig steam are made in separate fired boilers, similar calculations can be made. If 200 and 30 psig steam are obtained by reducing 400-psig steam, the 1737 Btu_{meth}/lb steam factor calculated for 400-psig steam also applies to the 200 and 30 psig steam.

The 400 psig steam produced by cogeneration (at a heat rate of 10,127 Btu_{meth}/kWh) has a methane equivalent of 1034.5 Btu_{meth}/lb, while the steam produced in a 70% efficient boiler has a methane equivalent of 1737.0 Btu_{meth}/lb. That is, 68% more fuel is required when 400-psig steam is made in a 70% efficient boiler. This comparison illustrates the fuel economy of cogeneration in producing steam efficiently.

There are some who believe that cogeneration makes electricity cheaper. But as the above calculations demonstrate, the real value of cogeneration is that it makes steam, not electricity, cheaper. That is because the alternative of making the same in a fired boiler is less efficient. There are numerous instances where corporate savings were unnoticed and unappreciated because an engineer failed to communicate and work with management.

20.3 Utilities Rate Structures

An understanding of utility rates (or tariffs) is important, because the cost of utility service is considered when evaluating a Building System Energy Economization (BSEE). Once the energy saving is determined, it should be given a dollar value.

The average cost of electricity does not accurately reflect the savings per kilowatt (kW) or kilowatt-hour (kWh). Virtually all electric utilities have general rules and regulations that include information regarding billing practices, available voltages, customer responsibilities, voltage regulation, balance and reliability, line extension limits, and temporary service requirements and availability. The rate schedules give the minimum (and maximum) values of usage to qualify for each rate along with the procedure for calculating the cost of electricity. Often, there are charges that apply to all rates, which are listed separately in another section and are called *riders*. Some common rates are residential, commercial, industrial, low-load factor, time of day, and area lighting. Some common riders are fuel-cost special voltages, and charges for extra facilities, such as redundant services or transformers. Reduced rates may be available for areas where the utility may have excess facilities (such as depressed areas) or where the customer undertakes serious building system demand-control implementations, such as thermal storage for air conditioning loads.

20.3.1 Billing Calculations

Each specific rate will contain a means to determine the cost for any or all of the following: kilowatt demand (kWd), kilowatt-hour (kWh), kilovolt ampere (kVA), kilovar (kvar), and power factor level. The charge may vary with time of day or time of year, and minimum service or customer charges may also be included. The billing kWd is the highest average rate of energy usage over the billing cycle. Usage (kWh) is generally averaged for a 15- or 30-minute demand period. The demand for a 15-minute period is determined by multiplying the kWh used in that period by 4.

Some electric rates contain a ratchet clause. This ratchet provision sets a minimum level for subsequent billings. This minimum usually continues for 3 to 11 months. The ratchet is generally applied only to the demand portion of the bill.

Declining Block Rate: To understand the use of the declining block rate let's assume that a building used 150,000 kWh and had a peak demand of 250 kWd. The first 50 kWd costs $5 per kWd or $250, leaving 200 kWd for the remaining blocks. The next block takes 150 kWd at $4.50 per kWd or $675, leaving only 50 kWd for the next block, which can accommodate 200. The last 50 kWd is then billed at $4 per kWd which costs $200. The total demand charge is the sum of the charges from each block, as follows:

Block 1	50 kWd − 5.0$/kWd	= $ 250
Block 2	150 kWd − 4.5$/kWd	= $ 675
Block 3	50 kWd − 4.0$/kWd	= $ 200
Total 250 kWd		$1125 or $4.50 kWd average

The charges for kWh may be performed in much the same manner as shown below:

$$kWh \times dollars/kWh = charge$$

Block 1	50,000 kWh × 0.007$/kWh = $3500	
Remaining = 150,000 − 50,000 = 100,000		
Block 2	50,000 × 0.065$/kWh	$3250
Remaining = 100,000 − 50,000 = 50,000		
Block 3	50,000 × 0.060$/kWh	$3000
Total	150,000 kWh for	$9750 or 6.5¢ per kWh average

The total bill is then $1125 + $9750, or $10,875. The average cost of electricity can be expressed in terms of kWd or kWh. The electric cost can be expressed as

43.50 dollars per kWd or 7.25 cents per kWh. It is important to note that averages cannot be used to determine energy savings. If the demand of this building is reduced by 25 kWd without an accompanying kWh reduction, the effect on the bill will be seen only in Block 3 of the demand charges (this rate of charge is referred to as the tail rate). The correct energy savings then is 25 × 4$/kWd or $100. By using the average total cost per kWd of $43.50, an erroneous savings of $1088 would be shown.

For purposes of economic evaluation of energy conservation only, the additional (rider) costs per kWh—particularly fuel costs (which may at any time represent a significant percentage of the basic energy cost)—may be estimated and added to the energy block rate. The utilities usually try to incorporate the major portion of fuel costs into the basic rates. Most rates include a charge for power factor by

- Assuming a power factor in the kWd charge;
- Charging for kVA;
- Charging for power factors below a given value; or
- Charging for kvars (reactive demand).

In any case, the utility is capturing its cost for supplying vars to its customers.

Demand Rate: A more complex rate is the demand usage block rate in which the sizes of certain kWh blocks are determined by the peak demand. This rate allows smaller consumers to take advantage of the lower kWh charge when their energy usage is high. (The load factor is a measure of energy usage relative to demand. This factor is the ratio of kWh to kWd times hours per billing cycle. Its value varies between the limits of 0 and 1.0). The demand usage rate allows a utility to reduce the number of rates and encourages a more consistent level of electric usage. The concept is best described by the following example.

Suppose that a building uses 500,000 kWh and has a peak demand of 1000 kWd. The kWh section of their rate schedule is as shown in the following data:

Blocks	Charge
(1) First 50,000 kWh	7 cents per kWh
(2) Next 200 kWh per kWd	6 cents per kWh
(3) Next 300 kWh per kWd	5 cents per kWh
(4) All excess	4 cents per kWh

Since building demand is 1000 kWd, the amount of kWh in Block 2 is 200 × 1000 or 200,000; similarly, the amount in Block 3 is 300,000. Once these block sizes are determined, the following procedure (identical to the previous example) can be used:

Block 1	50,000 × $0.07 =	$ 3500
(500,000 − 50,000 = 450,000 kWh remaining)		
Block 2	200,000 × $0.06 =	$12,000
(450,000 − 200,000 = 250,000 kWh remaining)		
Block 3	250,000 × $0.05 =	$12,500
Total 500,000 kWh		$28,000

The average cost is 5.6 cents per kWh.

This system of billing, in effect, combines the energy and demand charges. If needed for analysis purposes, this rate structure can be broken down into energy and demand components. Some simpler forms of this billing include the basic demand and energy structure with a reduced cost for all kWh over a designated number of hours of the maximum demand.

Time of Use Rate: The cost of new generating capacity for a utility is generally several times that of the cost of existing generating capacity. The difficulty in siting a new plant, and the long lead time for construction (typically 10 to 15 years) required for a new plant, together with the cost, may make it desirable for the utility to offer inducements for reduction of customer demand. Even for utilities with sufficient existing capacity, economic plant operating practice favors the highest load factors. The time-of-day utilizes different rate schedules for different times of day. "Peak" loads and "shoulder" loads (high but less than peak) would have higher cost rate schedules than those of night, weekend, or holiday loads.

Utility meters are available with three sets of registers that record the usage during the various periods. Electronic type meters permit even finer gradations. Some electronic meters can gather remote readings at a central location and accumulate data that will permit complex rate computations and/or usage analysis by computer. Utilities often offer lower rates for loads that are automatically controlled by the utility on a time-of-day basis, such as for water heating. Interruptible power rates may be of advantage to facilities where process control requirements permit the shutting down of loads.

20.4 Economics and Management of Energy

The thermal and other environmental control systems for buildings are already a major influence on construction costs, as well as on energy consumption in operating the building. As the techniques and equipment for reducing energy consumption are developed through such steps as recovery of waste heat, high-efficiency lighting, premium efficiency motors and transformers, and addition of variable frequency drives the percentage of mechanical and electrical system cost will probably become an even greater influence on the cost of constructing new buildings.

20.4.1   Energy Management

Any energy-conscious building engineer or facility manager should attempt to exert control over the building's energy usage (kWh) and the rate of usage (kWd). The simple fact is that no energy is used when equipment is shut off. Therefore, one of the first jobs of a building engineer is to make sure that unused, redundant, and idling equipment is shut off.

A further advantage to the user accrues from the fact that some utilities are offering their customers rebates that cover up to 40% of the cost of demand control and energy management equipment. The reasoning of the utility is simple: it presently costs $1500 ± per kilowatt of generating capacity for new power plant construction, whereas demand and energy management equipment costs only $80 ± per kilowatt demand reduction. Since a utility, by terms of its franchise, must supply all the power demanded, it is very much in the interest of any utility that is generating near capacity to reduce demand.

Most motor manufacturers now produce a line of high power factor, high-efficiency motors. The designer should match the motor as closely as possible to the load requirement and avoid the oversizing so common in practice. This will do much to improve power factor and efficiency and concomitantly to conserve energy. Where this is impractical because of variable loads, the variable frequency control device that matches input to load is the best choice. These devices reduce power frequency to reduce input when the motor is only partially loaded, thus reducing electrical and mechanical losses and increasing power factor.

20.4.2   Energy Economization

In existing facilities, before embarking on a building system energy economization program it is important to establish the existing pattern of energy usage and to identify areas where energy consumption could be reduced. A history of electric or gas usage, on a month-by-month basis, is available from electric or gas bills; this usage should be carefully recorded in a format (possibly graphic) that will facilitate future reference, evaluation, and analysis. The following is a basic list of items that should be recorded (where appropriate) in the electric or gas usage history (time of day rates may require multiple entries for usages and demands at each different time block for each month):

- Billing month
- Reading date
- Days in billing cycle
- Kilowatt-hours (or kilovolt ampere hours if billed on this basis)
- Billing kilowatt demand (or kilovolt ampere if billed on this basis)
- Actual kilowatt demand (or kilovolt ampere if billed on this basis)

- Kilovars (actual and billed)
- Kilovar hours (actual and billed)
- Power factor
- Power bill (broken down into the above categories plus fuel cost and any additional charges)
- Gas cubic foot or therms
- Heating or cooling degree days
- Additional column(s) for remarks (such as building occupancy periods).

A listing of building system operations and equipment will provide both a history and a basis for evaluating future improvement. The building system can be divided by function into five categories. The five functions are lighting, HVAC (heating, ventilating, and air conditioning) systems, motors and drives, electrical distribution equipment, and the building's environmental shell.

The listing of this information along with all types of energy usage is called an *energy audit.* In general, there are three categories of building system energy economization (BSEE) as follows:

1. *Housekeeping measures.* Easily performed (and usually low cost) actions that should logically be done (for example, turning lights off when not required; cleaning or changing air filters; cleaning heat exchangers; keeping doors shut; and shutting down redundant motors, pumps, and fans).
2. *Equipment modification.* This is usually more difficult and more expensive because it involves physical changes to the building system (for example, the addition of solid-state variable speed drives, reducing motor sizes on existing equipment, and modifying heating and cooling systems).
3. *Changes to the building shell.* Improving the insulating quality of the building to reduce losses to the outside environment (for example, adding insulation, reducing infiltration, controlling exhaust/intake, etc.).

It should be intuitively obvious that housekeeping and low-cost measures should be undertaken without delay. The larger and more expensive BSEE approaches generally take longer to initiate and get funded. However, there may be cases where obvious equipment modification improvements can be made concurrently with low-cost improvements. In some instances, constraints—particularly regarding energy availability—may require the expenditure of capital to increase existing energy efficiency to have sufficient capacity for future building expansion. This may occur where either utility capacity is severely limited or the costs associated with increasing service or pipe and wire size would be so high that expansions would essentially be made from current energy conservation load reductions.

If the process operation in the building produces waste steam then it can be used for building air conditioning (using a turbine refrigeration drive or absorption chillers), or for building heating. Building system energy balance is an important consideration. There may be cases where a BSEE action would merely shift the energy source from one point (or fuel) to another. For example, in an electrically heated building during winter, turning the lights off means more input to electric heaters, to make up heat that was supplied by the lights. In this case, the lights should have been left on, because the lighting is not only providing visibility but also adding free heat during the winter. On the other hand, lighting adds to the ac load in the summer so turning off lights reduces air conditioning and lighting power requirements. Hence, energy balance requires looking at the entire system or building shell.

20.4.3   Demand Control

The basic technique of demand control is a simple one; electric loads are disconnected and reconnected in such a fashion that demand peaks are leveled off and load factor thereby improved. The extent to which a facility's electric loads can tolerate this type of switching is an indication of the potential effectiveness of a demand controller. Installations with large uninterruptible loads, such as computers, will benefit minimally from demand control. Most industrial and commercial installations, however, contain a large percentage of interruptible loads (interruptions may be very short) and demand control systems frequently accomplish a 15% to 20% reduction in electric bills with a resultant short payback period on equipment investment.

The proliferation of demand control equipment has also produced a proliferation of nomenclature, including load shedding control, automated load control, peak demand control, and computerized load control. The last term refers to the type of equipment used to accomplish the intended control function with feedback from ambient lighting sensor, occupancy sensors, BTU meters, enthalpy controllers, etc. Descriptions that include the term "energy management" are inaccurate. Those devices are primarily concerned with control of *energy*. Demand control devices are intended to control *power demand,* with energy savings being an important but secondary benefit.

Demand controllers may operate to automatically reduce load, or they can be used to provide alarms to operating or supervising personnel for initiating action to reduce consumptions or demand. Semiautomatic controllers include time clocks that switch loads based on a predetermined time schedule. Demand controllers may be incorporated into the facility master control system. In this case, signals from photocells, environmentally controlled switches, and sensors have to be sent or telemetered to the control system. However, several considerations should be made when applying a controller. These considerations are as follows:

1. Each operation and piece of equipment shall be surveyed to find which loads can be switched off and to what extent they can be switched. The engineer shall evaluate any loss of equipment life or mechanical problems associated with switching each load. The survey, as a minimum, should list equipment by the following four categories: *Critical equipment.* Required at all times for production, safety, or other reasons. *Necessary.* This equipment can be shut down, if demand control gain exceed production loss. *Deferrable.* This equipment is important but can be turned off for varying periods of time. *Unnecessary.* This equipment has usually been left on, even though it is not needed. Sometimes equipment is used only occasionally and the user fails to de-energize after use, so an indicator or semiautomatic controller may be properly applied.

2. The operation of the equipment in an automatic mode should not endanger anyone near the equipment or inadvertently interrupt any process.

3. The controller should be periodically checked to see that it is operating as planned and has not been defeated.

4. In the cases of time clocks, the time should be checked and the time control adjusted to compensate for changing seasons and conditions.

5. The controlled equipment should be capable of withstanding the planned number of starts and stops. The controller should have a back-up power supply.

Demand controller or utility meter calculates the maximum demand by averaging the kWh over a set interval (a 15-minute demand interval would indicate the kWh for 15 minutes multiplied by 4 since there are four 15-minute periods per hour). The commonly used controller schemes are as follows:

1. *Instantaneous.* Controls loads at any time during an interval if the rate of usage exceeds a preset value.

2. *Ideal rate.* Controls loads when they exceed the set rate but allows a higher usage at the beginning of the interval.

3. *Converging rate.* Has a broader control bandwidth and an offset in the beginning of the interval but tightens control at the end of the interval.

4. *Predictive rate.* The controller is programmed to predict the usage at the end of the interval by the usage pattern along the interval and switches load to achieve the preset demand level.

5. *Continuous interval.* The controller looks at the past usage over a period equal to (or less than) the demand interval. Loads are switched in such a manner that no time period (of an interval's duration) will see an accumulation of kWh that exceeds the preset value. This controller needs no utility meter pulse.

20.5 Simplified Economic Analysis

Most engineers do not have an unlimited budget and, therefore, often need to make evaluations of various options. The energy savings of one or more projects have to be weighed against their "own-and-operate" costs. Since the equipment will usually function for many years and a future savings (or cost) is also involved, the time value of many should generally be considered.

To properly evaluate a BSEE, the installed cost as well as the operating, maintenance, and energy expenses should be determined on an annual basis over the life of equipment. Each annual expenditure (or savings) should be inflated and then discounted to the same base by the appropriate multiplier. If the engineer is unfamiliar with the process, he should at least develop the anticipated costs by year and work with the corporate accountant to determine the value of each option to the company.

Various levels of sophistication can be applied to economic analysis, depending on the desired accuracy of the results. One commonly used technique for evaluating systems relative to their projected savings is the *simple years to payback* approach. This approach is particularly useful for quick screening of alternatives. More detailed analysis, such as *present value* (also termed current value, current worth, or present worth) or *uniform annualized cost* analysis is usually warranted for the most attractive alternatives or when more complex time-dependent variations in maintenance requirements and other effects must be considered.

Simple Years to Payback: This simple technique ignores the cost of borrowing money (interest) and lost opportunity costs. It also ignores inflation and the time value of money.

Example 1. Equipment item 1 costs $10,000 and will save $2000 per year in operating costs, while equipment item 2 costs $12,000 and saves $3000 per year. Which item has the best simple payback?

$$\text{Item 1—}\$10,000/(\$2000/yr) = 5 \text{ year simple payback}$$
$$\text{Item 2—}\$12,000/(\$3000/yr) = 4 \text{ year simple payback}$$

Because the analysis of equipment for the duration of its realistic life can produce a much different result, the simple years to payback technique should be used with caution.

20.6   Sophisticated Economic Analysis

All sophisticated economic analysis methods use the basic principles of present value analysis to account for the time value of money. Therefore, a good understanding of these principles is important.

Single Payment Present Value Analysis: The cost or value of money is a function of the available interest rate and inflation rate. The future value F of a present sum of money P over n periods with compound interest rate i per period is

$$F = P (1 + i)^n \tag{1}$$

Conversely, the present value or present worth P of a future sum of money F is given by

$$P = F/(1 + i)^n \tag{2}$$

or

$$P = F \times \text{PWF} (i,n)_{\text{sgl}} \tag{3}$$

where the single payment present worth $\text{PWF}(i,n)_{\text{sgl}}$ is defined as

$$\text{PWF}(i,n)_{\text{sgl}} = 1/(1 + i)^n \tag{4}$$

Example 2. Calculate the future value of a building system presently valued at $10,000 in 10 years at 10% per year interest.

$$F = P (1 + i)^n = \$10{,}000 (1 + 0.1)^{10} = \$25{,}937.42$$

Series of Equal Payments: The present worth factor for a series of future equal payments (e.g., operating costs) is given by

$$\text{PWF}(i,n)_{\text{ser}} = \frac{(1+i)^n - 1}{i(1+i)^n} \tag{5}$$

The present value P of those future equal payments (PMT) is then the product of the present worth factor and the payment (i.e., $P = \text{PWF}(i,n)_{\text{sgl}} \times \text{PMT}$).

The future equal payments to repay a present value of money is determined by the capital recovery factor (CRF), which is the reciprocal of the present worth factor for a series of equal payments:

$$\text{CRF} = \text{PMT}/P \tag{6}$$

$$\text{CRF}(i,n) = \frac{(1+i)^n - 1}{i(1+i)^n} = \frac{i}{1 - (1+i)^n} \tag{7}$$

The capital recovery factor is often used to calculate periodic uniform loan payments.

When payment periods other than annual are to be studied, the interest rate must be expressed per appropriate period. For example, if monthly payments or returns on investment are being analyzed, then interest must be expressed per month, not per year, and n must be expressed in months, not years.

Example 3. Determine the present value of an annual operating cost of $1000/year over a 10-year period, assuming 10% per year interest rate.

$$PWF(i,n)_{ser} = [(1 + 0.1)^{10} - 1]/[0.1 (1 + 0.1)^{10}] = 6.14$$

$$Present\ value - \$1000\ (6.14) = \$6410$$

20.7 Microlevel Economic Analysis

This somewhat more sophisticated years to payback approach is similar to the simple years to payback method, except that the cost of money (interest rate, discount rate, etc.) is considered. Solving Eq. (7) for n yield the following:

$$n = \frac{\ln[CRF/(CRF - i)]}{\ln(1+i)} \tag{8}$$

Given known investment amounts and earnings, CRFs can be calculated for the alternative investments. Subsequently, the number of periods until payback has been achieved can be calculated using Eq. (8). Alternatively, a period-by-period (e.g., month-by-month or year-by-year) tabular cash flow analysis may be performed, or the necessary period to yield the calculated capital recovery factor may be obtained from a plot of CRFs.

Example 4. Compare the years to payback of the same items described in Example 1 if the value of money is 10% per year.

Item 1

    Cost = $10,000

  Savings = $2000/year

    CRF = $2000/$10,000 = 0.2

      $n = \ln[0.2/(0.2 - 0.1)]/\ln(1 + 0.1) = 7.3$ years

Item 2

    Cost = $12,000

  Savings = $3000/year

CRF = $3000/$12,000 = 0.25

$n = \ln[0.25/(0.25 - 0.1)]/\ln(1 + 0.1) = 5.4$ years

Accounting for Inflation: Different economic goods inflate at different rates. Inflation reflects the rise in the real cost of a commodity over time and is a separate issue from the time value of money. Inflation must often be accounted for in an economic evaluation. One way to account for inflation is to substitute effective interest rates that account for inflation into the equations given here.

The effective interest rate i', sometimes called the real rate, accounts for inflation rate j and interest rate i or discount rate i_d; it can be expressed as follows:

$$i' = \frac{1+i}{1+j} - 1 = \frac{i-j}{1+j} \tag{9}$$

Example 5. Determine the present worth P of an annual operating cost of $1000 over 10 years, given a discount rate of 10% per year and an inflation rate of 5% per year.

$$i' = (0.1 - 0.05)/(1 + 0.05) = 0.0476$$

$$\text{PWF}(i',n)_{\text{ser}} = \frac{(1+0.0476)^{10} - 1}{0.0476(1+0.0476)^{10}} = 7.813$$

$$P = \$1000 \,(7.813) = \$7813$$

The most frequently used economic analysis techniques are to examine all costs and incomes to be incurred over the analysis period (1) in terms of their present value (i.e., today's or initial year's value, also called constant value); (2) in terms of equal periodic costs or payments (uniform annualized costs); or (3) in terms of periodic cash flows (e.g., monthly or annual cash flows). Each method provides a slightly different insight. The present value methods allows easy comparison of alternatives over the analysis period chosen. The uniform annualized costs method allows comparison of average annual costs of different options. The cash flow method allows comparison of actual cash flows rather than average cash flows; it can identify periods of overall positive and negative cash flow, which is helpful for cash management purposes.

Computer analysis software such as the Life Cycle Program eases calculations and performs parametric or sensitivity analyses. Some commercial spreadsheet programs include economic analysis functions. However, any of the analyses should consider more details of both positive and negative costs over the analysis period, such as varying inflation rates, capital and interest costs, salvage costs, replacement costs, interest deductions, depreciation allowances, taxes, tax credits, mortgage payments, and all other costs associated with a particular system. The replacement cost should be evaluated after checking service life of equipment. *Table 20.3* provides estimated service life for major building system M/E equipment.

Sophisticated Present Value Method: The total present value (present worth) for any analysis is determined by summing the present worths of all individual items under consideration, both future single payment items and series of equal future payments, as described earlier. The scenario with the highest present value is the preferred alternative.

Uniform Annualized Costs Methods: It is sometimes useful to project a uniform periodic (e.g., average annual) cost over the analysis period. The basic procedure for determining uniform annualized costs is to first determine the present worth of all costs and then apply the capital recovery factor to determine equal payments over the analysis period.

Uniform annualized building system owning, operating, and maintenance costs can be expressed, as

$$C_y = - \text{ capital and interest} + \text{salvage value} - \text{replacements}$$
$$- \text{ disposals} - \text{operating energy} - \text{property tax}$$
$$- \text{ maintenance} - \text{insurance} + \text{interest tax deduction}$$
$$+ \text{ depreciation}$$

where

$$\text{Capital and interest} = (C_{s,init} - \text{ITC}) \, \text{CRF}(i', n) \text{ where}$$
$$\text{ITC} = \text{investment tax credit}$$
$$C_{s,init} = \text{system initial cost}$$

$$\text{Salvage value} = C_{s,salv} \, \text{PWF}(i', n)\text{CRF}(i', n)(1 - T_{salv}) \text{ where}$$
$$T_{salv} = \text{salvage tax rate} \quad \text{and}$$
$$C_{s,salv} = \text{system salvage value}$$

$$\text{Replacements or disposals} = \sum_{k=1}^{m} [R_k \text{PWF}(i',k)]\text{CRF}(i',n)(1 - T_{inc})$$

$$\text{Operating energy} = C_e[\text{CRF}(i', n)/\text{CRF}(i'', n)](1 - T_{inc})$$

$$\text{Property tax} = C_{s,assess}T_{prop}(1 - T_{inc})$$

$$\text{Maintenance} = M(1 - T_{inc})$$

$$\text{Insurance} = I(1 - T_{inc})$$

The equations and formula's appearing on pages 335 to 340 were reprinted with permission of the American Society of Heating, Refrigerating and Airconditioning Engineers, Atlanta, Georgia, from the 1995 ASHRAE Handbook—Applications.

Table 20.3 Estimates of Service Lives of Major Building System Components

Equipment Item	Median Years	Equipment Item	Median Years
Air conditioners		Electric motors	18
Window unit	10	Motor starters	17
Residential single or split package	15	Electrical transformers	30
Commercial through-the-wall	15	Controls	
Water-cooled package	15	Pneumatic	20
Heat pumps		Electric	16
Residential air-to-air	15	Electronic	15
Commercial air-to-air	15	Valve actuators	
Commercial water-to-air	19	Hydraulic	15
Rooftop air conditioners		Pneumatic	20
Single-zone	15	Self-contained	10
Multizone	15	Air terminals	
Boilers, hot water (steam)		Diffusers, grilles, and registers	27
Steel water-tube	24(30)	Induction and fan-coil units	20
Steel fire-tube	25(25)	VAV and double-duct boxes	20
Cast iron	35(30)	Air washers	17
Electric	15	Ductwork	30
Burners	21	Dampers	20
Furnaces		Fans	
Gas- or oil-fired	18	Centrifugal	25
Unit heaters		Axial	20
Gas or electric	13	Propeller	15
Hot water or steam	20	Ventilating roof-mounted	20
Radiant heaters		Coils	
Electric	10	DX, water, or steam	20
Hot water or steam	25	Electric	15
Air-cooled condensers	20	Heat exchangers	
Evaporative condensers	20	Shell-and-tube	24
Insulation		Reciprocating compressors	20
Molded	20	Package chillers	
Blanket	24	Reciprocating	20
Pumps		Centrifugal	23
Base-mounted	20	Absorption	23
Pipe-mounted	10	Cooling towers	
Sump and well	10	Galvanized metal	20
Condensate	15	Wood	20
Reciprocating engines	20	Ceramic	34
Steam turbines	30		

$$\text{Interest tax deduction} = T_{\text{inc}} \sum_{k=1}^{n} [i_{\text{m}} P_{k-1,i} \text{PWF}(i_{\text{d}}, k)] \text{CRF}(i', n)$$

$$\text{Depreciation} = T_{\text{inc}} \sum_{k=1}^{n} [D_k \text{PWF}(i_{\text{d}}, k)] \text{CRF}(i', n)$$

The outstanding principle P_k during year k at market rate i_{m} is given by

$$P_k = (C_{\text{s,init}} - \text{ITC}) \left[(1 + i_{\text{m}})^{k-1} + \frac{(1 + i_{\text{m}})^{k-1} - 1}{(1 + i_{\text{m}})^{-n} - 1} \right] \qquad (10)$$

Note: P_k is in current dollars and must, therefore, be discounted by the discount rate i_{d}, not i'. Likewise, the summation term for interest deduction can be expressed as

$$\sum_{k=1}^{n} [i_{\text{m}} P_k / (1 + i_{\text{d}})^k] = (C_{\text{s,init}} - \text{ITC})$$

$$\times \left[\frac{\text{CRF}(i_{\text{m}}, n)}{\text{CRF}(i_{\text{d}}, n)} + \frac{1}{(i + i_{\text{m}})} \frac{i_{\text{m}} - \text{CRF}(i_{\text{m}}, n)}{\text{CRF}[i_{\text{d}} - i_{\text{m}}) / (1 + i_{\text{m}}), n]} \right] \qquad (11)$$

If $i_{\text{d}} = i_{\text{m}}$,

$$\sum_{k=1}^{n} [i_{\text{m}} P_k / (1 + i_{\text{d}})^k] = (C_{\text{s,init}} - \text{ITC}) \times \left[1 + \frac{n}{(1 + i_{\text{m}})} [i_{\text{m}} - \text{CRF}(i_{\text{m}}, n)] \right] \qquad (12)$$

Depreciation terms commonly used include depreciation calculated by the straight line depreciation method, which is

$$D_{k,\text{SL}} = (C_{\text{s,init}} - C_{\text{s,salv}}) / n$$

and the sum-of-digits depreciation method:

$$D_{k,\text{SD}} = (C_{\text{s,init}} - C_{\text{s,salv}}) [2(n - k + 1)] / n(n + 1)$$

The following example illustrates the use of the uniform annualized cost method.

Example 6. Calculate the annualized system cost using constant dollars for a $10,000 system considering the following factors: a 5-year life, a salvage value of $1000 at the end of the 5 ears, no investment tax credits, a $500 replacement

in year 3, a discount rate i_d of 10%, a general inflation rate j of 5%, a fuel inflation rate j_e of 8%, a market mortgage rate i_m of 10%, an annual operating cost for
energy of $500, a $100 annual maintenance cost, a $50 annual insurance cost,
straight line depreciation, an income tax rate of 50%, a property tax rate of 1% of
assessed value, an assessed system value equal to 40% of the initial system
value, and a salvage tax rate of 50%.

Effective interest rate i' for all but fuel:

$$i' = (i_d - j)/(1 + j) = (0.10 - 0.05)/(1 + 0.05) = 0.047619$$

Effective interest rate i'' for fuel:

$$i'' = (i_d - j_e)/(1 + j_e) = (0.10 - 0.08)/(1 + 0.08) = 0.018519$$

Capital recovery factor CRF(i', n) for items other than fuel:

$$\text{CRF}(i', n) = i'/[1 - (1 + i')^{-n}] = 0.47619/[1 - (1.047619)^{-5}] = 0.229457$$

Capital recovery factory CRF(i'', n) for fuel

$$\text{CRF}(i'', n) = i''/[1 - (1 + i'')^{-n}] = 0.018519/[1 - (0.018519)^{-5}] = 0.211247$$

Capital recovery factory CRF(i_m, n) for loan or mortgage

$$\text{CRF}(i_m, n) = i_m/[1 - (1 + i_m)^{-n}] = 0.01/[1 - (1.10)^{-5}] = 0.263797$$

Loan payment = $10,000 (0.263797) = $2637.97

Present worth factor PWF(i_d, years 1 to 5)

$$\text{PWF}(i_d, 1) = 1/1.10)^1 = 0.909091$$

$$\text{PWF}(i_d, 2) = 1/1.10)^2 = 0.826446$$

$$\text{PWF}(i_d, 3) = 1/1.10)^3 = 0.751315$$

$$\text{PWF}(i_d, 4) = 1/1.10)^4 = 0.683013$$

$$\text{PWF}(i_d, 5) = 1/1.10)^5 = 0.620921$$

Present worth factor PWF(i', years 1 to 5)

$$\mathrm{PWF}(i', 1) = 1/(1.047619)^1 = 0.954545$$

$$\mathrm{PWF}(i', 2) = 1/(1.047619)^2 = 0.911157$$

$$\mathrm{PWF}(i', 3) = 1/(1.047619)^3 = 0.869741$$

$$\mathrm{PWF}(i', 4) = 1/(1.047619)^4 = 0.830207$$

$$\mathrm{PWF}(i', 5) = 1/(1.047619)^5 = 0.792471$$

Capital and interest

$$(C_{s,init} - ITC)\,\mathrm{CRF}(i', n) = (\$10{,}000 - \$0)\,0.229457 = \$2294.57$$

Salvage value

$$C_{s,salv}\mathrm{PWF}(i', n)\mathrm{CRF}(i', n)\,(1 - T_{salv}) = \$1000 \times 0.792471 \times 0.229457 \times 0.5$$
$$= \$90.92$$

Replacements

$$\sum_{k=1}^{m}[R_k\mathrm{PWF}(i', k)]\mathrm{CRF}(i', n)(1 - T_{inc}) = \$500 \times 0.869741 \times 0.229457 \times 0.5$$

$$= \$49.89$$

Operating energy

$$C_e[\mathrm{CRF}(i', n)/\mathrm{CRF}(i'', n)]\,(1 - T_{inc}) = 500\,[0.229457/0.211247]\,0.5 = \$271.55$$

Property tax

$$C_{s,assess}T_{prop}(1 - T_{inc}) = \$10{,}000 \times 0.40 \times 0.01 \times 0.5 = \$20.00$$

Maintenance

$$M(1 - T_{inc}) = \$100\,(1 - 0.5) = \$50.00$$

Insurance

$$i(1 - T_{inc}) = \$50\,(1 - 0.5) = \$25.00$$

Interest deduction

$$T_{inc}\sum_{k=1}^{n}[i_m P_{k-1}\text{PWF}(i_d,k)]\text{CRF}(i',n)$$

Table 20.2 summarizes the interest and principle payments for this example. Annual payments are the product of the initial system cost $C_{s,init}$ and the capital recovery factor $\text{CRF}(i_m, 5)$. Also, Eq. (10) can be used to calculate total discounted interest deduction directly.

Next, apply the capital recovery factor $\text{CRF}(i', 5)$ an tax rate T_{inc} to the total of the discounted interest sum.

$$\$2554.66\ \text{CRF}(i', 5)T_{inc} = \$2554.66 \times 0.229457 \times 0.5 = \$293.09$$

Depreciation

$$T_{inc}\sum_{k=1}^{n}[D_{k,SL}\text{PWF}(i_d,k)]\text{CRF}(i',n)...$$

Use the straight line depreciation method to calculate depreciation:

$$D_{k,SL} = (C_{s,init} - C_{s,salv})/n = (\$10,000 - \$1000)/5 = \$1800.00$$

Next, discount the depreciation:

Year	$D_{k,SL}$	PWF(i_d, k)	Discounted depreciation
1	$1800.00	0.909091	$1636.36
2	$1800.00	0.826446	$1487.60
3	$1800.00	0.751315	$1352.37
4	$1800.00	0.683013	$1229.42
5	$1800.00	0.620921	$1117.66
		Total	$6823.42

Finally, the capital recovery factor and tax are applied.

$$\$6823.42\ \text{CRF}(i', n)T_{inc} = \$6823.42 \times 0.229457 \times 0.5 = \$782.84$$

The U.S. tax code recommends estimating the salvage value prior to depreciating; for other countries use local tax code suggestions. Then depreciation is claimed as the difference between the initial and the salvage value, which is the way depreciation is treated in this example. The more common practice is to ini-

Table 20.2 Interest Deduction Summary (for Example 6)

Year	Payment Amount, Current $	Interest Payment, Current $	Principal Payment, Current $	Outstanding Principal, Current $	PWF(i_d, k)	Discounted Interest, Discounted $	Discounted Payment, Discounted $
0	—	—	—	10,000.00	—	—	—
1	2637.97	1000.00	1637.97	8362.02	0.909091	909.09	2398.17
2	2637.97	836.20	1801.77	6560.26	0.826446	691.07	2180.14
3	2637.97	656.03	1981.95	4578.31	0.751315	492.89	1981.95
4	2637.97	457.83	2180.14	2398.17	0.683013	312.70	1801.77
5	2637.97	239.82	2398.17	0	0.620921	148.91	1637.97
Total	—	3189.88	10,000.00	—	—	2554.66	10,000.00

tially claim zero salvage value, and at the end of ownership of the item, treat any salvage value as a capital gain.

Summary of Terms

Capital and interest	−$ 2294.57
Salvage value	+$ 90.92
Replacements	−$ 49.89
Operating costs	−$ 271.55
Property tax	−$ 20.00
Maintenance	−$ 50.00
Insurance	−$ 25.00
Interest deduction	+$ 293.00
Depreciation deduction	+$ 782.84
Total annualized cost	−$ 1544.16

20.8   Building System Equipment Service Life and Impact on Buildings

To provide true life cycle cost analysis all areas impacted by system selection need to be included in the study model. Table 20.4 shows some of the other areas of building impacted by building system components, arrangements, and performance requirements.

Table 20.4 Project and Building Elements Impacted by Building System Selection

Areas Impacted by Building System Selection	Conceptual	Schematic	Design Development
General Project Budget Layout Criteria & Standards	—Design concepts —Program interpretation —Site resources —Site system components Circulation —Project budget —Design intentions	—Schematic floor plans —Schematic sections —Approach to systems —Integration —Floor to floor height —Functional space —Envelope relationships	—Floor plans —Sections —Typical details —Integrated systems —Space circulation —Specifications —Site resources integration
Structural Integration Foundation Substructure Superstructure	—Performance requirements for Building systems —Structural sizing by system need —Framing systems integration —Subsurface system layout —Underground system concepts	—Schematic site plan —Selection of foundation based on system location and need —Structural system integration and selection	—Key foundation Elements, details covering system concepts —Sizing of major system Support elements —Outline specifications
Architectural Integration Envelop Exterior closure Roofing Interior construction Elevators Equipment	—Approach to elevations —Views to/from building —Interior design —Configuration of key spaces —Impact of key equipment on facility & site —Passive solar usage —Passive geothermal usage —Active thermal storage	—Concept elaboration —Selection of envelope systems —Schematic elevations —Selection of sound attenuation —System equipment and architecture Aesthetics —Basic elevator & vertical transportation concepts —Impact of key system equipment on space design	—Key elevation details showing system component elevations —Key roofing details showing system components —Interior system Elements —Integration of structural framing

Table 20.4 (*continued*)

Areas Impacted by Building System Selection	Conceptual	Schematic	Design Development
Building Systems HVAC Plumbing Fire Protection Service & distribution Lighting & power Building automation	—Basic energy concepts —Impact of passive/active system Concepts on facility —Initial systems selection —Space allocation —Performance requirements for plumbing, HVAC, fire protection —Basic power supply —Approaches to use of natural & artificial lighting —Performance requirements for lighting —Need for special electrical systems Systems integration and automation	—Systems selection —Refinement of service & distribution concepts —Input to schematic plans —Energy conservation —Windows/skylight Design & sizing Selection of lighting & electrical systems —General service, power & distribution concepts	—Detailed system selection —Initial system drawings & key details —Distribution & riser diagrams —Outline specifications for system elements —Detailed systems selection —Distribution diagrams —Key space Lighting layouts —Outline specification for electrical elements
Site and System Integration Preparation Utilities Landscaping System/integration Site/system scaping	—Site selection —Site resources utilization criteria —Site forms & massing —Requirements for system equipment access —Views to/from facility —Utility supply —Site drainage	—Design concept Elaboration —Initial site plan —Schematic planting, grading, paving plans	—Site plan —Planting plan —Typical site details —Outline specification for site materials

21

Building System and Construction Documents

21.1 Drawings

Building system designers will usually be given preliminary architectural drawings as a first step. These drawings permit the designers to arrive at the preliminary scope of the work, roughly estimate the requirements, and determine in a preliminary way the location of equipment and the methods of heating/ventilating and cooling and types of lighting. In this stage of the design, such items as primary and secondary distribution systems and major items of equipment will be decided. The early requirements for types of machinery to be installed will be determined.

Early in the design period, the designer should emphasize the need for room to hang conduits pipes, ducts and cable trays, crawl spaces, structural reinforcements for equipment, and special floor loadings; and for clearances around cooling/heating equipment, substations, switchgear, transformers, panelboards, switchboards, and other items that may be required. It is much more difficult to obtain such special requirements once the design has been committed. The need for installing, removing, and relocating machinery must also be considered.

21.2 Specifications

A contract for installation of building electrical/mechanical system requires both a written document and drawings. The written document contains both legal (non-technical) and engineering (technical) sections. The legal section contains the general terms of the agreement between contractor and owner, such as payment, working conditions, and time requirements, and it may include clauses on performance bonds, extra work, penalty clauses, and damages for breach of contract.

The engineering section includes the technical specifications. The specifications give descriptions of the work to be done and the materials to be used. It is common practice in larger installations to use a standard outline format listing division, section, and subsection titles or subjects of the Construction Specifications Institute (CSI) (see Figure 21.1) Where several specialities are involved, Division 16 covers the electrical/mechanical installation and Division 15 covers the mechanical portion of the work. The building or plant automation system, integrating several building control systems, is covered in CSI Division 13—Special Construction. It is important to note that some electrical/mechanical work will almost always be included in CSI Divisions 13(thirteen) and 1(one).

To assist the engineer in preparing contract specifications, standard technical specifications (covering construction, application, technical, and installation details) are available from technical publishers and manufacturers (which may require revision to avoid proprietary specifications). Large organizations, such as the U.S. Government General Services Administration and the Veterans Administration, develop their own standard specifications. The engineer should keep several cautions in mind when using standard specifications. First, they are designed to cover a wide variety of situations, and consequently they will contain considerable material that will not apply to the specific facility under consideration, and they may lack other material that should be included. Therefore, standard specifications must be appropriately edited and supplemented to embody the engineer's intentions fully and accurately. Second, many standard specifications contain material primarily for nonindustrial facilities and may not reflect the requirements of the specific industrial processes.

Computer-aided specifications (CAS) have been developed that will automatically create specifications as an output from the CAE-CADD process.

The following section provides a uniform approach to organizing specification text. It consist of three Parts: General, Products, and Execution.

PART 1: GENERAL

SUMMARY
   Section Includes
   Products supplied but not installed Under This Section
   Products Installed But Not Supplied Under This Section
   Related Sections
   Allowances
   Unit Prices
   Measurement Procedures
   Payment Procedures
   Alternates/Alternatives*

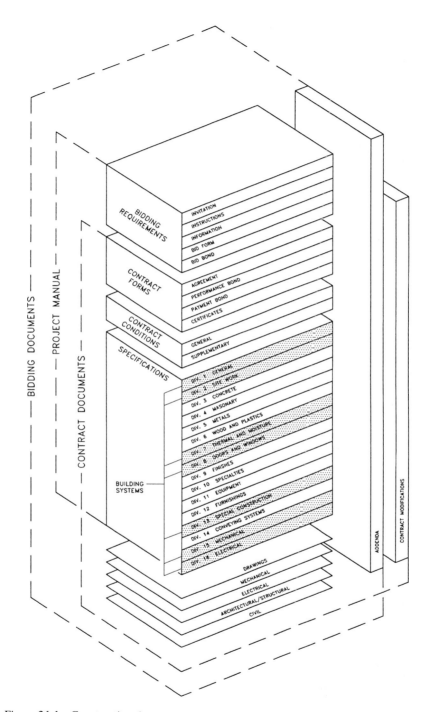

Figure 21.1 Construction documents.

REFERENCES

DEFINITIONS

SYSTEM DESCRIPTION
  Design Requirements
  Performance Requirements

SUBMITTALS
  Product Data
  Shop Drawings
  Samples
  Quality
  Assurance/Control Submittals
    Design Data,
    Test Reports,
    Certificates,
    Manufacturer's Instructions,
    Manufacturer's Field Reports
  Closeout Submittals

QUALITY ASSURANCE
  Qualifications
  Regulatory Requirements
  Certifications
  Field Samples
  Mockups
  Pre-Installation Meetings

DELIVERY, STORAGE, AND HANDLING
  Packing, Shipping, Handling, and Unloading
  Acceptance at Site Storage and Protection

PROJECT CONDITIONS OR SITE CONDITIONS*
  Environmental Requirements
  Existing Conditions

SEQUENCING

SCHEDULING

WARRANTY
  Special Warranty

SYSTEM STARTUP

OWNER'S INSTRUCTIONS

COMMISSIONING

MAINTENANCE
Extra Materials
Maintenance Service

PART 2: PRODUCTS

MANUFACTURERS

EXISTING PRODUCTS

MATERIALS

MANUFACTURED UNITS

EQUIPMENT

COMPONENTS

ACCESSORIES

MIXES

FABRICATION
Shop Assembly

FINISHES
Shop Printing
Shop Finishing

SOURCE QUALITY CONTROL
Fabrication Tolerances Tests,
Inspection Verification of Performance

PART 3: EXECUTION

ACCEPTABLE INSTALLERS

EXAMINATION
Site Verification of Conditions

PREPARATION
Protection Surface Preparation

ERECTION

INSTALLATION

APPLICATION

CONSTRUCTION
  Special Techniques
  Interface with Other Work
  Sequence of Operation
  Site Tolerances

REPAIR/RESTORATION

RE-INSTALLATION

FIELD QUALITY CONTROL
  Site Tests
  Inspection
  Manufacturer's Field Services

ADJUSTING

CLEANING

DEMONSTRATION

PROTECTION

SCHEDULES

The following is a list of basic building system materials, equipment, and products assembled in construction specification institute format. See Figures 21–1 and 21–2 for integrated building system specification blocks.

Division - 1
01650 FACILITY STARTUP/COMMISSIONING
  -655 Starting of System
  -660 Testing, Adjusting, and Balancing of Systems
  -670 Systems demonstrations

Division - 2
02600 UTILITY PIPING MATERIALS
  -605 Utility Structures
          Cleanouts
          Manholes and Covers
          Tunnels
  -610 Pipe and Fittings
          Cast Iron Pipe
          Concrete Pipe
          Corrugated Metal Pipe
          Ductile Iron Pipe
          Mineral Fiber Reinforced Cement
          Pipe

Plastic Pipe
Pre-Insulated Pipe
Steel Pipe
Vitrified Clay Pipe
-640 Valves and Cocks
-645 Hydrants

02660 WATER DISTRIBUTION
-665 Water Systems
Chilled Water Systems
Cisterns
Domestic Water Systems
Fire Water Systems
Heating Water Systems
Thrust Restraints
-670 Water Wells
Well Drilling and Casing
-675 Disinfection of Water Distribution Systems

02680 FUEL AND STEAM DISTRIBUTION
-685 Gas Distribution Systems
-690 Oil Distribution Systems
Fuel Tanks
-695 Steam Distribution Systems

02780 POWER AND COMMUNICATIONS
-785 Electric Power Transmission
Overhead Electric Power
Transmission
Underground Electric Power
Transmission
-790 Communication Transmission
Fiber Optics Communications
Microwave Communications
Shortwave Communications
Satellite Antennas

Division - 5
05900 HYDRAULIC STRUCTURES
-910 Penstocks
-915 Bulkheads
-920 Trashracks
-925 Manifolds
-930 Bifurcations

Division - 7

07190 VAPOR RETARDERS

07195 AIR BARRIERS

07200 INSULATION
- -210 Building Insulation
    - Batt Insulation
    - Building Board Insulation
    - Foamed-in-place Insulation
    - Loose Fill Insulation
- -220 Roof and Deck Insulation
    - Asphaltic Perlite Concrete Deck
    - Roof Board Insulation

07800 SKYLIGHTS
- -810 Plastic Unit Skylights
- -820 Metal Framed Skylights

Division - 8

08650 SPECIAL WINDOWS
- -655 Roof Windows
- -660 Security Windows and Screens
- -665 Pass Windows
- -670 Storm Windows

08700 HARDWARE
- -740 Electro-Mechanical Hardware
- -760 Window Hardware
    - Automatic Window Equipment
    - Window Operators
    - Window Locks and Lifts
- -770 Door And Window Accessories

08800 GLAZING
- -810 Glass
    - Float Glass
    - Rolled Glass
    - Tempered Glass
    - Laminated Glass
    - Insulating Glass
    - Coated Glass
    - Mirrored Glass
    - Wired Glass
- -840 Plastic Glazing
    - Insulating Plastic Glazing
- -850 Glazing Accessories

Division - 10

10200 LOUVERS AND VENTS
  -210 Metal Wall Louvers
        Operable Metal Wall Louvers
        Stationary Metal Wall Louvers
        Motorized Metal Wall Louvers
  -220 Louvered Equipment Enclosures
  -225 Metal Door Louvers
  -230 Metal Vents
        Metal Soffit Vents
        Metal Wall Vents

10520 FIRE PROTECTION SPECIAL-TIES
  -522 Fire Extinguishers, Cabinets, and Accessories
        Fire Extinguishers
        Fire Extinguisher Cabinets
  -526 Fire Blankets and Cabinets
  -528 Wheeled Fire Extinguish Units

Division - 13

13010 AIR SUPPORTED STRUCTURES

13020 INTEGRATED ASSEMBLIES
  -025 Integrated Ceilings
  -027 Integrated Access Floors

13600 SOLAR ENERGY SYSTEMS
  -610 Solar Flat Plate Collectors
        Air Collectors
        Liquid Collectors
  -620 Solar concentrating Collectors
  -625 Solar Vacuum Tube Collectors
  -630 Solar Collector Components
        Solar Absorber Plates and Tubing
        Solar Coatings and Surface
      Treatment
        Solar Collector Insulation
        Solar Glazing
        Solar Housing and Framing
        Solar Reflectors
  -640 Packaged Solar Systems
  -650 Photovoltaic Collectors

13700 WIND ENERGY SYSTEMS

13750 COGENERATION SYSTEMS

13800 BUILDING AUTOMATION SYSTEMS
 -810 Energy Monitoring and Control Systems
 -815 Environmental Control Systems
 -820 Communications Systems
 -825 Security Systems
 -830 Clock Control Systems
 -835 Elevator Monitoring and Control Systems
 -840 Escalators and Moving Walks Monitoring and Control Systems
 -845 Alarm and Detection Systems
 -850 Door Control Systems

13900 FIRE SUPPRESSION AND SUPERVISORY SYSTEMS

13950 SPECIAL SECURITY CONSTRUCTION
>   Crash Barriers
>   Deal Drawers
>   Gun Ports
>   Money Cart Pass-Through
>   Security Door Frame Protection
>   Security Gates

Division - 15
15050 BASIC MECHANICAL MATERIALS AND METHODS
 -060 Pipes and Pipe Fittings
>   Aluminum and Aluminum Alloy
>   Pipe and Fittings
>   Concrete Pipe and Fittings
>   Copper and Copper Alloy Pipe and
>   Fittings
>   Ferrous Pipe and Fittings
>   Fiber Pipe and Fittings
>   Glass Pipe and Fittings
>   Hoses and Fittings
>   Plastic Pipe and Fittings
>   Pre-Insulated Pipe and Fittings
 -100 Valves
>   Manual Control Valves
>   Self Actuated Valves
 -120 Piping Specialties
 -130 Gages
 -140 Supports and Anchors
 -150 Meters
 -160 Pumps
 -170 Motors

-175 Tanks
-190 Mechanical Identification
-240 Mechanical Sound, Vibration, and Seismic Control

15250 MECHANICAL INSULATION
-260 Piping Insulation
-280 Equipment Insulation
-290 Ductwork Insulation

15300 FIRE PROTECTION
-310 Fire Protection Piping
-320 Fire Pumps
-330 Wet Pipe Sprinkler Systems
-335 Dry Pipe Sprinkler Systems
-340 Pre-Action Sprinkler Systems
-345 Combination Dry Pipe and Pre-Action Sprinkler Systems
-350 Deluge Sprinkler Systems
-355 Foam Extinguishing Systems
-360 Carbon Dioxide Extinguishing Systems
-370 Dry Chemical Extinguishing Systems
-375 Standpipe and Hose Systems

15400 PLUMBING
-410 Plumbing Piping
-430 Plumbing Specialties
-440 Plumbing Fixtures
-450 Plumbing Equipment
       Domestic Water Heat Exchangers
       Drinking Water Cooling Systems
       Pumps
       Storage Tanks
       Water Conditioners
       Water Filtration Devices
       Water Heaters
-475 Pool and Fountain Equipment
-480 Special Systems
       Compressed Air Systems
       Deionized Water Systems
       Distilled Water Systems
       Fuel Oil Systems
       Gasoline Dispensing Systems
       Helium Gas Systems
       Liquefied Petroleum Gas Systems
       Lubricating Oil Systems

Natural Gas Systems
Nitrous Oxide Gas Systems
Oxygen Gas Systems
Reverse Osmosis Systems
Vacuum Systems

15500 HEATING, VENTILATING, AND AIR CONDITIONING
-510 Hydronic Piping
-515 Hydronic Specialties
-520 Steam and Steam Condensate Piping
-525 Steam and Steam Condensate Specialties
-530 Refrigerant Piping
-535 Refrigerant Specialties
-540 HVAC Pumps
-545 Chemical Water Treatment

15550 HEAT GENERATION
-555 Boilers
-570 Boiler Accessories
-575 Breechings, Chimneys, and Stacks
-580 Feedwater Equipment
-590 Fuel Handling Systems
-610 Furnances
-620 Fuel Fired Heaters
     Duct Furnaces
     Gas Fired Unit Heaters
     Oil Fired Heaters
     Radiant Heaters

15650 REFRIGERATION
-655 Refrigeration Compressors
-670 Condensing Units
-680 Water Chillers
     Absorption Water Chillers
     Centrifugal Water Chillers
     Reciprocating Water Chillers
     Rotary Water Chillers
-710 Cooling Towers
     Mechanical Draft Cooling Towers
     Natural Draft Cooling Towers
-730 Liquid Coolers
-740 Condensers

15750 HEAT TRANSFER
-755 Heat Exchangers

-760 Energy Storage Tanks
-770 Heat Pumps
　　　　Air Source Heat Pumps
　　　　Rooftop Heat Pumps
　　　　Water Source Heat Pumps
-780 Packaged Air Conditioning Units
　　　　Computer Room Air Conditioning Units
　　　　Packaged Rooftop Air Conditioning Units
　　　　Packaged Terminal Air Conditioning Units
　　　　Unit Air Conditioners
-790 Air Coils
-810 Humidifiers
-820 Dehumidifiers
-830 Terminal Heat Transfer Units
　　　　Convectors
　　　　Fan Coil Units
　　　　Finned Tube Radiation
　　　　Induction Units
　　　　Unit Heaters
　　　　Unit Ventilators
-845 Energy Recovery Units

15850 AIR HANDLING
-855 Air Handling Units with Coils
-860 Centrifugal Fans
-865 Axial Fans
-870 Power Ventilators
-875 Air Curtain Units

15880 AIR DISTRIBUTION
-885 Air Cleaning Devices
　　　　Dust Collectors
　　　　Filters
-890 Ductwork
　　　　Metal Ductwork
　　　　Nonmetal Ductwork
　　　　Flexible Ductwork
　　　　Ductwork Hangars and Supports
-910 Ductwork Accessories
　　　　Dampers
　　　　Duct Access Panels and Test Holes
　　　　Duct Connection Systems
　　　　Flexible Duct Connections
　　　　Turning Vanes and Extractors

-920 Sound Attenuators
-930 Air Terminal Units
   Constant Volume
   Variable Volume
-940 Air Outlets and Inlets
   Diffusers
   Intake and Relief Ventilators
   Louvers
   Registers and Grilles

15950 CONTROLS
-955 Building Systems Control
-960 Energy Management and Conservation Systems
-970 Control Systems
   Electric Control Systems
   Electronic Control Systems
   Pneumatic Control Systems
   Self-Powered Control Systems
-980 Instrumentation
-985 Sequence of Operation

15990 TESTING, ADJUSTING, AND BALANCING
-991 Mechanical Equipment Testing, Adjusting, and Balancing
-992 Piping Systems Testing, Adjusting, and Balancing
-993 Air Systems Testing, Adjusting, and Balancing
-994 Demonstration of Mechanical Equipment
-995 Mechanical System Start-up/Commissioning

Division - 16
16050 BASIC ELECTRICAL MATERIALS AND METHODS
-110 Raceways
   Cable Trays
   Conduits
   Surface Raceways
   Indoor Service Poles
   Underground Ducts
   Underground Ducts and Manholes
-120 Wires and Cables
   Fiber Optic Cable
   Low Voltage Wire
   600 Volt or Less Wire and Cable
   Medium Voltage Cable
   Undercarpet Cable Systems

-130 Boxes
  Floor Boxes
  Outlet Boxes
  Pull and Junction Boxes
-140 Wiring Devices
  Low Voltage Switching
-150 Manufactured Wiring Systems
-160 Cabinets and Enclosures
-190 Supporting Devices
-195 Electrical Identification

16200 POWER GENERATION - BUILT-UP SYSTEMS
-210 Generators
  Solar Electric Generators
  Steam Electric Generators
-250 Generator Controls
  Instrumentation
  Starting Equipment
-290 Generator Grounding

16300 MEDIUM VOLTAGE DISTRIBUTION
-310 Medium Voltage Substations
-320 Medium Voltage Transformers
-330 Medium Voltage Power Factor Correction
-340 Medium Voltage Insulators and Lightning Arrestors
-345 Medium Voltage Switchboard
-350 Medium Voltage Circuit Breakers
-355 Medium Voltage Reclosers
-360 Medium Voltage Interrupter Switches
-365 Medium Voltage Fuses
-370 Medium Voltage Overhead Power Distribution
-375 Medium Voltage Underground Power Distribution
-380 Medium Voltage Converters
  Medium Voltage Frequency
Changers
  Medium Voltage Rectifiers
-390 Medium Voltage Primary Grounding

16400 SERVICE AND DISTRIBUTION
-410 Power Factor Correction
-415 Voltage Regulators
-420 Service Entrance

-425 Switchboards
-430 Metering
-435 Converters
-440 Disconnect Switches
-445 Peak Load Controllers
-450 Secondary Grounding
-460 Transformers
-465 Bus Duct
-470 Panelboards
      Branch Circuit Panelboards
      Distribution Panelboards
-475 Overcurrent Protective Devices
      Circuit Breakers
      Fuses
-480 Motor Control
-485 Contactors
-490 Switches
      Transfer Switches
      Isolation Switches

16500 LIGHTING
-501 Lamps
-502 Luminaire Accessories
      Ballasts
      Lenses
      Lighting Maintenance Equipment
      Light Louvers
      Posts and Standards
-510 Interior Luminaires
      Fluorescent Luminaires
      High Intensity Discharge
      Luminaires
      Incandescent Luminaires
      Luminous Ceilings
-520 Exterior Luminaires
      Aviation Lighting
      Flooding Lighting
      Navigation Lighting
      Roadway Lighting
      Signal Lighting
      Site Lighting
      Sport Lighting
-535 Emergency Lighting

-545 Underwater Lighting
-580 Theatrical Lighting

16600 SPECIAL SYSTEMS
-610 Uninterruptible Power Supply Systems
-620 Packaged Engine Generator Systems
-630 Battery Power Systems
      Central Battery Systems
      Packaged Battery Systems
-640 Cathodic Protection
-650 Electromagnetic Shielding Systems
-670 Lightning Protection Systems
-680 Unit Power Conditioners

16700 COMMUNICATIONS
-720 Alarm and Detection Systems
      Fire alarm Systems
      Smoke Detection Systems
      Gas Detection Systems
      Intrusion Detection Systems
      Security Access Systems
-730 Clock and Program Systems
-740 Voice and Data Systems
      Telephone Systems
      Paging System
      Call System
      Data Systems
      Local Area Network Systems
      Door Answering Systems
      Microwave and Radio Systems
      Central Dictation Systems
      Intercommunication Systems
-770 Public Address and Music Systems
-780 Television Systems
      Master Antenna Systems
      Video Telecommunication Systems
      Video Surveillance Systems
      Broadcast Video Systems
-785 Satellite Earth Station Systems
-790 Microwave Systems

16850 ELECTRIC RESISTANCE HEATING
-855 Electric Heating Cables and Mats
-880 Electric Radiant Heaters

16900 CONTROLS
  -910 Electrical Systems Control
  -915 Lighting Control Systems
        Dimming Systems
  -920 Environmental Systems Control
  -930 Building Systems Control
  -940 Instrumentation

16950 TESTING
  -960 Electrical Equipment Testing
  -970 Electrical System Startup/Commissioning
  -980 Demonstration of Electrical Equipment

21.3 Design Considerations and Coordination

Electrical/mechanical equipment usually occupies a relatively small percentage of the total building space and, in design, it may be easier to relocate electrical/mechanical service areas than structural elements. Allocation of space for electrical/mechanical areas is often given secondary consideration. In the competing search for space, the building system engineer is responsible for fulfilling the requirements for a proper electrical/mechanical installation while recognizing the flexibility of electrical/mechanical systems in terms of layout and placement.

Depending on the type and complexity of the project, the M/E engineer will need to co-ordinate with a variety of other specialists, the minimum number of specification divisions and sections that require co-ordination is shown in figure 21.2. These potentially include chemical, process, civil, structural, industrial, production, lighting, fire protection, and environmental engineers; maintenance planners; architects; representatives of federal, state, and local regulatory agencies; interior and landscape designers; specification writers; construction and installation contractors; lawyers; purchasing agents; applications engineers from major equipment suppliers and the local electrical/mechanical utility; and management staff of the organization that will operate the facility.

The building system designer must become familiar with local rules and know the authorities having jurisdiction over the design and construction. It can be inconvenient and embarrassing to have an electrical/mechanical project held up at the last moment because proper permits have not been obtained; for example, a permit for a street closing to allow installation of utilities to the site or an environmental permit for an on-site generator.

Local contractors are usually familiar with local ordinances and union work rules and can be of great help in avoiding pitfalls. In performing electrical/mechanical design, it is essential, at the outset, to prepare a checklist of all the de-

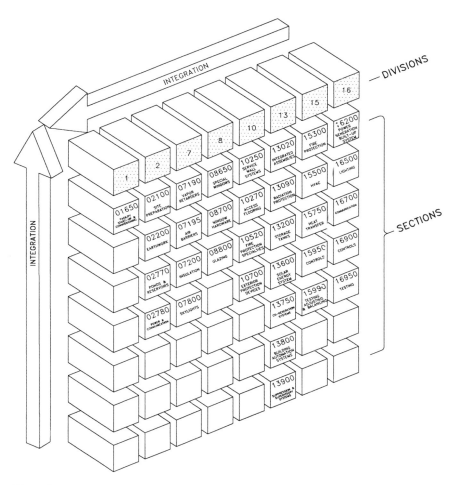

Figure 21.2   Integrated building system specifications.

sign stages that have to be considered. Major items include temporary power, access to the site, and review by others. Certain electrical/mechanical work may appear in nonelectrical/mechanical sections of the specifications. For example, furnishing and connecting of electric motors and motor controllers may be covered in the mechanical section of the specifications. For administrative control purposes, the electrical/mechanical work may be divided into a number of contracts, some of which may be under the control of a general contractor and some of which may be awarded to electrical/mechanical contractors.

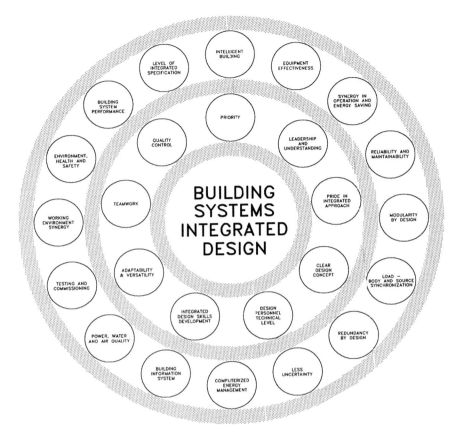

Figure 21.3 Requirements for optimized integration.

21.4 Integrated Design Integrals

To integrate all tangibles of a project require all the rollers of bearing shown in Figure 21.3.

Following areas of project should be overemphasized for integration of all disciplines.

Safety: When design engineers lay out equipment rooms and locate electrical/mechanical equipment, they cannot always avoid having some areas accessible to unqualified persons. Dead-front construction should be utilized whenever practical. Proper barricading, signing, and guarding should be installed and maintained on energized systems or around machinery that could be hazardous, or that is located in occupied areas. Work rules, especially in areas of medium or high voltage electrical equipment, should be established.

Work on energized power systems or equipment should be permitted only where qualified staff is available to perform such work and only if it is essential. This is foremost a matter of safety, but is also required to prevent damage to equipment. A serious cause of failure, attributable to human error, is unintentional grounding or phase-to-phase short circuiting of equipment that is being worked on. By careful design, such as proper spacing and barriers, and by enforcement of published work-safety rules, the designer can minimize this hazard. Unanticipated backfeeds through control circuitry, from capacitors, instrument transformers, or test equipment, presents a danger to the worker.

Protective devices, such as ground-fault relays and ground-fault detectors (for high-resistance or ungrounded systems), will minimize damage from electrical/mechanical failures. Electrical fire and smoke can cause maintenance staff to disconnect all electric power, even if there is no direct danger to the occupants. Electrical/mechanical failures that involve smoke and noise, even though occurring in unoccupied areas, may cause confusion to the working population. Nuisance tripping, which may interrupt industrial processes, can be minimized by careful design and selection of protective equipment.

Maintenance: This is essential to proper operation. The installation should be so designed that maintenance can be performed with normally available maintenance personnel (either in-house or contract). Design details should provide proper space, accessibility, and working conditions so that the systems can be maintained without difficulty and excessive cost.

Generally, the external electrical systems are operated and maintained by the electrical utility, though at times they are a part of the building distribution system. Where continuity of service is essential, suitable transfer equipment and alternate sources should be provided. Such equipment is needed to maintain minimum lighting requirements for passageways, stairways, and critical areas as well as to supply power to critical loads. These systems usually include automatic or manual equipment for transferring loads on loss of normal supply power or for putting battery or generator-fed equipment into service.

Flexibility: Flexibility of the building system means adaptability to development and expansion as well as to changes to meet varied requirements during the life of the facility. Sometimes a designer is faced with providing power in a building where the loads may be unknown. For example, some manufacturing buildings are constructed with the occupied space designs incomplete (shell and core designs). In some cases, the designer will provide only the core utilities available for connection by others to serve the working areas. In other cases, the designer may lay out only the basic systems and, as the tenant requirements are developed, fill in the details. A manufacturing division or tenant may provide working space designs.

Because it is usually difficult and costly to increase the capacity of feeders, it is important that provisions for sufficient capacity be provided initially. Modern facilities, including manufacturing, may require frequent relocations of equipment, addition of production equipment, function modifications, and even movement of equipment to and from other sites; therefore, a high degree of system flexibility is an important design consideration. Extra conductors or duct/piping space should be included in the design stage when additional loads are added. When required, space must be provided for outdoor substations, underground system including spare ducts, and overhead distribution.

Flexibility in a building system is enhanced by the use of oversize or spare raceways, cables, busways, manifold piping/ducting, isolation valves/dampers, and equipment. The cost of making such provisions is usually relatively small in the initial installation. Space on spare support hangers and openings (sealed until needed) between walls and floors may be provided at relatively low cost for future work. Openings through floors should be sealed with fireproof (removable) materials to prevent the spread of fire and smoke between floors. For computer rooms and similar areas, flexibility is frequently provided by raised floors made of removable panels, providing access to a wiring space between the raised floor and the slab below. Similarly in Industrial Buildings for greater economy and flexibility, Plug-in busways and trolley-type busways provide a convenient method of serving machinery subject to relocation.

Politics: This involves getting benefits, usually competitively. Within an organization the principal benefits sought are: raises, promotions, budgets, perks, power (breadth and depth of authority), ego, and prestige. This cultivates some classes of behavior such as:

1. *Defensiveness.* Some people feel insecure if they are found to be in error and fight desperately to prove that they are not in error.

2. *Competitiveness.* If controlled, competitors can be great achievers; if uncontrolled, they can be merely destructive and obstructive.

3. *Conceit.* Whatever is done, it must show how great I am.

4. *Negativeness.* Sometimes a masked hostility.

5. *Stubbornness.* Devotion to the first idea.

The building system engineer's entire professional life is a battle with Murphy's Law: "If something can go wrong, it will." We design organized systems out of less organized structures and components, and we design these systems to overcome failures. To do system design a thorough review of overall picture from the outside in and from the inside out is required. With mutual respect it is constructive, as in brainstorming sessions, but with tender egos, conflicting political motives, and disrespect, it generates quarrels. System design concepts require continuous adjustment until the freeze bell rings.

One can recite a variety of *Rules for Building System Design,* but most are obvious, such as the "rule" that one should study the owner's specification before starting to design. Most owner specifications are not written with integration in mind, because they are produced by individuals who have always worked in their own discipline considering everybody outside that discipline boundary inferior. So the only serious "rule" is, "It depends." Here are a few principles that may help you:

1. Try to change one thing at a time since many of the changes will be debugging the bugs you put in to overcome the original bugs. Measure the ego temperature toward cooperation and do not proceed until you understand ego and correct its cause.

2. Nobody likes construction change orders. In choosing design alternatives, predictability is a value. A reliably predictable design may be better than a more elegant design whose performance you cannot reliably predict.

3. Provide quantitative specifications if you can. Qualitative requirements cause misunderstandings and change orders when acceptance time comes around.

But if you integrate well, you may see the results in the actual operation of the building.

21.5 Preliminary Cost Estimate

A preliminary estimate is usually requested. Sometimes the nature of a preliminary estimate makes it nothing more than a good guess. Enough information is usually available, however, to perform the estimate on a square foot, per process machine, per production area, by the horsepower or number of motors, or on a similar basis for a comparable facility.

A second estimate is often provided after the project has been clearly defined but before any drawings have been prepared. The electrical/mechanical designer can determine from sketches and architectural layouts the type of lighting fixtures as well as many items of heavy equipment that are to be used. Lighting fixtures, as well as most items of heavy equipment, can be priced directly from the catalogs, using appropriate discounts.

The most accurate estimate is made when drawings have been completed and bids are about to be received or the contract negotiated. The estimating procedure of the designer in this case is similar to that of the contractor's estimator. It involves first the takeoffs, that is, counting the number of receptacles, lighting fixtures, lengths of wire and conduit, determining the number and types of equipment, and then applying unit costs for labor, materials, overhead, and profit.

368 / Integrated Mechanical/Electrical Design

21.6 Shop Drawings

After the design has been completed and contracts are awarded, contractors, manufacturers, and other suppliers will submit drawings for review or information. It is important to review and comment upon these drawings and return them as quickly as possible; otherwise, the supplier and/or contractor may claim that the work was delayed by the engineer's review process. Unless the drawings contain serious errors and/or omissions, it is usually a good practice not to reject them but to stamp the drawings with terminology such as "revise as noted" and mark them to show errors, required changes, and corrections. The supplier can then make appropriate changes and proceed with the work without waiting to re-submit the drawings for approval.

If the shop drawings contain major errors or discrepancies, however, they should be rejected with a requirement that they be resubmitted to reflect appropriate changes that are required on the basis of notes and comments of the engineer.

Unless otherwise directed, communication with contractors and suppliers is always through the construction (after inspection) authority. In returning corrected shop drawings, remember that the contract for supplying the equipment is usually with the general contractor and that the official chain of communication is through him or her. Sometimes direct communication with a subcontractor or a manufacturer may be permitted; however, the content of such communication should always be confirmed in writing with the general contractor. Recent lawsuits have resulted in placing the responsibility for shop drawing correctness (in those cases and possibly future cases) upon the design engineer, leaving no doubt that checking is an important job.

Index